Free Study Tips DVD

In addition to the tips and content in this guide, we have created a FREE DVD with helpful study tips to further assist your exam preparation. **This FREE Study Tips DVD provides you with top-notch tips to conquer your exam and reach your goals.**

Our simple request in exchange for the strategy-packed DVD is that you email us your feedback about our study guide. We would love to hear what you thought about the guide, and we welcome any and all feedback—positive, negative, or neutral. It is our #1 goal to provide you with top-quality products and customer service.

To receive your **FREE Study Tips DVD**, email freedvd@apexprep.com. Please put "FREE DVD" in the subject line and put the following in the email:

> a. The name of the study guide you purchased.
>
> b. Your rating of the study guide on a scale of 1-5, with 5 being the highest score.
>
> c. Any thoughts or feedback about your study guide.
>
> d. Your first and last name and your mailing address, so we know where to send your free DVD!

Thank you!

ISEE Lower Level Test Prep

ISEE Study Guide and Practice Exam Questions [2nd Edition Book]

Matthew Lanni

ISEE Lower Level Test Prep

ISEE Study Guide and Practice Exam Questions [2nd Edition Book]

Matthew Lanni

Written and edited by APEX Publishing.

ISBN 13: 9781628457124
ISBN 10: 1628457120

APEX Publishing is not connected with or endorsed by any official testing organization. APEX Publishing creates and publishes unofficial educational products. All test and organization names are trademarks of their respective owners.

The material in this publication is included for utilitarian purposes only and does not constitute an endorsement by APEX Publishing of any particular point of view.

For additional information or for bulk orders, contact info@apexprep.com.

Table of Contents

Test Taking Strategies

1. Reading the Whole Question

A popular assumption in Western culture is the idea that we don't have enough time for anything. We speed while driving to work, we want to read an assignment for class as quickly as possible, or we want the line in the supermarket to dwindle faster. However, speeding through such events robs us from being able to thoroughly appreciate and understand what's happening around us. While taking a timed test, the feeling one might have while reading a question is to find the correct answer as quickly as possible. Although pace is important, don't let it deter you from reading the whole question. Test writers know how to subtly change a test question toward the end in various ways, such as adding a negative or changing focus. If the question has a passage, carefully read the whole passage as well before moving on to the questions. This will help you process the information in the passage rather than worrying about the questions you've just read and where to find them. A thorough understanding of the passage or question is an important way for test takers to be able to succeed on an exam.

2. Examining Every Answer Choice

Let's say we're at the market buying apples. The first apple we see on top of the heap may *look* like the best apple, but if we turn it over we can see bruising on the skin. We must examine several apples before deciding which apple is the best. Finding the correct answer choice is like finding the best apple. Although it's tempting to choose an answer that seems correct at first without reading the others, it's important to read each answer choice thoroughly before making a final decision on the answer. The aim of a test writer might be to get as close as possible to the correct answer, so watch out for subtle words that may indicate an answer is incorrect. Once the correct answer choice is selected, read the question again and the answer in response to make sure all your bases are covered.

3. Eliminating Wrong Answer Choices

Sometimes we become paralyzed when we are confronted with too many choices. Which frozen yogurt flavor is the tastiest? Which pair of shoes look the best with this outfit? What type of car will fill my needs as a consumer? If you are unsure of which answer would be the best to choose, it may help to use process of elimination. We use "filtering" all the time on sites such as eBay® or Craigslist® to eliminate the ads that are not right for us. We can do the same thing on an exam. Process of elimination is crossing out the answer choices we know for sure are wrong and leaving the ones that might be correct. It may help to cover up the incorrect answer choice. Covering incorrect choices is a psychological act that alleviates stress due to the brain being exposed to a smaller amount of information. Choosing between two answer choices is much easier than choosing between all of them, and you have a better chance of selecting the correct answer if you have less to focus on.

4. Sticking to the World of the Question

When we are attempting to answer questions, our minds will often wander away from the question and what it is asking. We begin to see answer choices that are true in the real world instead of true in the world of the question. It may be helpful to think of each test question as its own little world. This world may be different from ours. This world may know as a truth that the chicken came before the egg or may assert that two plus two equals five. Remember that, no matter what hypothetical nonsense may be in the question, assume it to be true. If the question states that the chicken came before the egg, then choose your answer based on that truth. Sticking to the world of the question means placing all of our biases and assumptions aside and relying on the question to guide us to the correct answer. If we are simply looking for answers that are correct based on our own judgment, then we may choose incorrectly. Remember an answer that is true does not necessarily answer the question.

5. Key Words

If you come across a complex test question that you have to read over and over again, try pulling out some key words from the question in order to understand what exactly it is asking. Key words may be words that surround the question, such as *main idea, analogous, parallel, resembles, structured,* or *defines*. The question may be asking for the main idea, or it may be asking you to define something. Deconstructing the sentence may also be helpful in making the question simpler before trying to answer it. This means taking the sentence apart and obtaining meaning in pieces, or separating the question from the foundation of the question. For example, let's look at this question:

> Given the author's description of the content of paleontology in the first paragraph, which of the following is most parallel to what it taught?

The question asks which one of the answers most *parallels* the following information: The *description* of paleontology in the first paragraph. The first step would be to see *how* paleontology is described in the first paragraph. Then, we would find an answer choice that parallels that description. The question seems complex at first, but after we deconstruct it, the answer becomes much more attainable.

6. Subtle Negatives

Negative words in question stems will be words such as *not, but, neither,* or *except*. Test writers often use these words in order to trick unsuspecting test takers into selecting the wrong answer—or, at least, to test their reading comprehension of the question. Many exams will feature the negative words in all caps (*which of the following is NOT an example*), but some questions will add the negative word seamlessly into the sentence. The following is an example of a subtle negative used in a question stem:

> According to the passage, which of the following is *not* considered to be an example of paleontology?

If we rush through the exam, we might skip that tiny word, *not*, inside the question, and choose an answer that is opposite of the correct choice. Again, it's important to read the question fully, and double check for any words that may negate the statement in any way.

7. Spotting the Hedges

The word "hedging" refers to language that remains vague or avoids absolute terminology. Absolute terminology consists of words like *always, never, all, every, just, only, none,* and *must*. Hedging refers to words like *seem, tend, might, most, some, sometimes, perhaps, possibly, probability,* and *often*. In some cases, we want to choose answer choices that use hedging and avoid answer choices that use absolute terminology. It's important to pay attention to what subject you are on and adjust your response accordingly.

8. Restating to Understand

Every now and then we come across questions that we don't understand. The language may be too complex, or the question is structured in a way that is meant to confuse the test taker. When you come across a question like this, it may be worth your time to rewrite or restate the question in your own words in order to understand it better. For example, let's look at the following complicated question:

> Which of the following words, if substituted for the word *parochial* in the first paragraph, would LEAST change the meaning of the sentence?

Let's restate the question in order to understand it better. We know that they want the word *parochial* replaced. We also know that this new word would "least" or "not" change the meaning of the sentence. Now let's try the sentence again:

> Which word could we replace with *parochial*, and it would not change the meaning?

Restating it this way, we see that the question is asking for a synonym. Now, let's restate the question so we can answer it better:

> Which word is a synonym for the word *parochial*?

Before we even look at the answer choices, we have a simpler, restated version of a complicated question.

9. Predicting the Answer

After you read the question, try predicting the answer *before* reading the answer choices. By formulating an answer in your mind, you will be less likely to be distracted by any wrong answer choices. Using predictions will also help you feel more confident in the answer choice you select. Once you've chosen your answer, go back and reread the question and answer choices to make sure you have the best fit. If you have no idea what the answer may be for a particular question, forego using this strategy.

10. Avoiding Patterns

One popular myth in grade school relating to standardized testing is that test writers will often put multiple-choice answers in patterns. A runoff example of this kind of thinking is that the most common answer choice is "C," with "B" following close behind. Or, some will advocate certain made-up word patterns that simply do not exist. Test writers do not arrange their correct answer choices in any kind of pattern; their choices are randomized. There may even be times where the correct answer choice will be the same letter for two or three questions in a row, but we have no way of knowing when or if this might happen. Instead of trying to figure out what choice the test writer probably set as being correct, focus on what the *best answer choice* would be out of the answers you are presented with. Use the tips above, general knowledge, and reading comprehension skills in order to best answer the question, rather than looking for patterns that do not exist.

FREE DVD OFFER

Achieving a high score on your exam depends not only on understanding the content, but also on understanding how to apply your knowledge and your command of test taking strategies. **Because your success is our primary goal, we offer a FREE Study Tips DVD. It provides top-notch test taking strategies to help you optimize your testing experience.**

Our simple request in exchange for the strategy-packed DVD is that you email us your feedback about our study guide.

To receive your **FREE Study Tips DVD**, email freedvd@apexprep.com. Please put "FREE DVD" in the subject line and put the following in the email:

   a. The name of the study guide you purchased.

   b. Your rating of the study guide on a scale of 1-5, with 5 being the highest score.

   c. Any thoughts or feedback about your study guide.

   d. Your first and last name and your mailing address, so we know where to send your free DVD!

Introduction to the ISEE Lower Level

Function of the Test

The Independent School Entrance Exam (ISEE) Lower Level is developed by the Educational Records Bureau (ERB) for student admissions into independent schools all over the world. Independent schools use the ISEE score as one of the determinants for entrance into their school, although other factors are considered as well, so check with your school to find out the weight of test scores in determining entrance. The ISEE Lower Level is for students entering into the 5th or 6th grade.

Test Administration

The ISEE Lower Level is offered in the United States and abroad in Prometric testing centers, at ISEE test site schools, and ISEE testing offices. The exam is offered within three testing seasons: Fall (August through November), Winter (December through March), and Spring/Summer (April through July). Students can take the exam up to three times within one 12-month admission cycle, but no more than once per season. Students with physical challenges or documented learning differences may register for special accommodations at the www.iseetest.org website.

Test Format

Students should remember to bring their verification letter or ID when they enter the testing premises. Inside the testing room, items like scratch paper, calculators, watches, rulers, protractors, dictionaries, cell phones, electronic devices, and thesauruses are not permitted. Students will receive two ten-minute breaks: one after the Quantitative Reasoning section, and the other following the Math Achievement section.

The ISEE Lower Level is offered as a computer-based test or a pencil and paper test. The exam has five sections: Verbal Reasoning, Quantitative Reasoning, Reading Comprehension, Mathematics Achievement, and an Essay. Below is more information on the sections:

Section	Questions	Time
Verbal Reasoning	34	20 minutes
Quantitative Reasoning	38	35 minutes
Reading Comprehension	25	25 minutes
Mathematics Achievement	30	30 minutes
Essay	1	30 minutes
	Total:	**140 minutes**

Scoring

For the exam, students should make an educated guess if they are not sure of the answer because there is no guessing penalty. Scores are based on the number of choices answered correctly, and points are not taken away for incorrect answer choices. The raw score is how many questions students get correct out of the number of questions answered. A scaled score is a raw score converted into a number between 760 and 940. A percentile score is also provided in the score report, with sometimes a stanine score given, which is a number given in place of percentile rank. Every school is different in determining the value of the scores. Students are compared to those in the same grade for the past 3 years.

Recent Developments

Retesting rules changed after the year 2016. Students now have the option to take the ISEE 3 times during a calendar year, whereas before they only had one opportunity to take the ISEE.

Verbal Reasoning

Synonyms

Analyzing Word Parts

By learning some of the etymologies of words and their parts, readers can break down new words into components and analyze their combined meanings. For example, the root word *soph* is Greek for wise or knowledge. Knowing this informs the meanings of English words including *sophomore, sophisticated,* and *philosophy.* Those who also know that *phil* is Greek for love will realize that *philosophy* means the love of knowledge. They can then extend this knowledge of *phil* to understand *philanthropist* (one who loves people), *bibliophile* (book lover), *philharmonic* (loving harmony), *hydrophilic* (water-loving), and so on. In addition, *phob-* derives from the Greek *phobos,* meaning fear. This informs all words ending with it as meaning fear of various things: *acrophobia* (fear of heights), *arachnophobia* (fear of spiders), *claustrophobia* (fear of enclosed spaces), *ergophobia* (fear of work), and *hydrophobia* (fear of water), among others.

Some English word origins from other languages, like ancient Greek, are found in large numbers and varieties of English words. An advantage of the shared ancestry of these words is that once readers recognize the meanings of some Greek words or word roots, they can determine or at least get an idea of what many different English words mean. As an example, the Greek word *métron* means to measure, a measure, or something used to measure; the English word meter derives from it. Knowing this informs many other English words, including *altimeter, barometer, diameter, hexameter, isometric,* and *metric.* While readers must know the meanings of the other parts of these words to decipher their meaning fully, they already have an idea that they are all related in some way to measures or measuring.

While all English words ultimately derive from a proto-language known as Indo-European, many of them historically came into the developing English vocabulary later, from sources like the ancient Greeks' language, the Latin used throughout Europe and much of the Middle East during the reign of the Roman Empire, and the Anglo-Saxon languages used by England's early tribes. In addition to classic revivals and native foundations by the Renaissance era, other influences included French, German, Italian, and Spanish. Today we can often discern English word meanings by knowing common roots and affixes, particularly from Greek and Latin.

The following is a list of common prefixes and their meanings:

Prefix	Definition	Examples
a-	without	atheist, agnostic
ad-	to, toward	advance
ante-	before	antecedent, antedate
anti-	opposing	antipathy, antidote
auto-	self	autonomy, autobiography
bene-	well, good	benefit, benefactor
bi-	two	bisect, biennial
bio-	life	biology, biosphere
chron-	time	chronometer, synchronize
circum-	around	circumspect, circumference

com-	with, together	commotion, complicate
contra-	against, opposing	contradict, contravene
cred-	belief, trust	credible, credit
de-	from	depart
dem-	people	demographics, democracy
dis-	away, off, down, not	dissent, disappear
equi-	equal, equally	equivalent
ex-	former, out of	extract
for-	away, off, from	forget, forswear
fore-	before, previous	foretell, forefathers
homo-	same, equal	homogenized
hyper-	excessive, over	hypercritical, hypertension
in-	in, into	intrude, invade
inter-	among, between	intercede, interrupt
mal-	bad, poorly, not	malfunction
micr-	small	microbe, microscope
mis-	bad, poorly, not	misspell, misfire
mono-	one, single	monogamy, monologue
mor-	die, death	mortality, mortuary
neo-	new	neolithic, neoconservative
non-	not	nonentity, nonsense
omni-	all, everywhere	omniscient
over-	above	overbearing
pan-	all, entire	panorama, pandemonium
para-	beside, beyond	parallel, paradox
phil-	love, affection	philosophy, philanthropic
poly-	many	polymorphous, polygamous
pre-	before, previous	prevent, preclude
prim-	first, early	primitive, primary
pro-	forward, in place of	propel, pronoun
re-	back, backward, again	revoke, recur
sub-	under, beneath	subjugate, substitute
super-	above, extra	supersede, supernumerary
trans-	across, beyond, over	transact, transport
ultra-	beyond, excessively	ultramodern, ultrasonic, ultraviolet
un-	not, reverse of	unhappy, unlock
vis-	to see	visage, visible

The following is a list of common suffixes and their meanings:

Suffix	Definition	Examples
-able	likely, able to	capable, tolerable
-ance	act, condition	acceptance, vigilance
-ard	one that does excessively	drunkard, wizard
-ation	action, state	occupation, starvation
-cy	state, condition	accuracy, captaincy
-er	one who does	teacher
-esce	become, grow, continue	convalesce, acquiesce
-esque	in the style of, like	picturesque, grotesque
-ess	feminine	waitress, lioness
-ful	full of, marked by	thankful, zestful
-ible	able, fit	edible, possible, divisible
-ion	action, result, state	union, fusion
-ish	suggesting, like	churlish, childish
-ism	act, manner, doctrine	barbarism, socialism
-ist	doer, believer	monopolist, socialist
-ition	action, result, state,	sedition, expedition
-ity	quality, condition	acidity, civility
-ize	cause to be, treat with	sterilize, mechanize, criticize
-less	lacking, without	hopeless, countless
-like	like, similar	childlike, dreamlike
-ly	like, of the nature of	friendly, positively
-ment	means, result, action	refreshment, disappointment
-ness	quality, state	greatness, tallness
-or	doer, office, action	juror, elevator, honor
-ous	marked by, given to	religious, riotous
-some	apt to, showing	tiresome, lonesome
-th	act, state, quality	warmth, width
-ty	quality, state	enmity, activity

Meanings of Words, Sentences, and Entire Texts

When readers encounter an unfamiliar word in a text, they can use the surrounding context, including the overall subject matter, specific chapter/section topic, and the immediate sentence context. Among others, one category of context clues is grammar. For example, the position of a word in a sentence and its relationship to the other words can help the reader establish whether the unfamiliar word is a verb, a noun, an adjective, or an adverb. This narrows down the possible meanings of the word to one part of speech. However, this may be insufficient. In a sentence that many birds *migrate* twice yearly, the reader can determine the word is a verb, and probably does not mean eat or drink; but it could mean travel, mate, lay eggs, hatch, molt, etc.

Some words can have a number of different meanings depending on how they are used. For example, the word *fly* has a different meaning in each of the following sentences:

- "His trousers have a fly on them."
- "He swatted the fly on his trousers."
- "Those are some fly trousers."
- "They went fly fishing."
- "She hates to fly."
- "If humans were meant to fly, they would have wings."

As strategies, readers can try substituting a familiar word for an unfamiliar one and see whether it makes sense in the sentence. They can also identify other words in a sentence, offering clues to an unfamiliar word's meaning.

Sentence Completion

Context Clues

Readers can often figure out what unfamiliar words mean without interrupting their reading to look them up in dictionaries by examining context. Context includes the other words or sentences in a passage. One common context clue is the root word and any affixes (prefixes/suffixes). Another common context clue is a synonym or definition included in the sentence. Sometimes both exist in the same sentence. Here's an example:

Scientists who study birds are *ornithologists*.

Many readers may not know the word *ornithologist*. However, the example contains a definition (scientists who study birds). The reader may also have the ability to analyze the suffix (*-logy*, meaning the study of) and root (*ornitho-,* meaning bird).

Another common context clue is a sentence that shows differences. Here's an example:

Birds *incubate* their eggs outside of their bodies, unlike mammals.

Some readers may be unfamiliar with the word *incubate*. However, since we know that "unlike mammals," birds incubate their eggs outside of their bodies, we can infer that *incubate* has something to do with keeping eggs warm outside the body until they are hatched.

In addition to analyzing the etymology of a word's root and affixes and extrapolating word meaning from sentences that contrast an unknown word with an antonym, readers can also determine word meanings from sentence context clues based on logic. Here's an example:

Birds are always looking out for predators that could attack their young.

The reader who is unfamiliar with the word *predator* could determine from the context of the sentence that predators usually prey upon baby birds and possibly other young animals. Readers might also use the context clue of etymology here, as *predator* and *prey* have the same root.

Denotation and Connotation

Denotation refers to a word's explicit definition, like that found in the dictionary. Denotation is often set in comparison to connotation. **Connotation** is the emotional, cultural, social, or personal implication associated with a word. Denotation is more of an objective definition, whereas connotation can be more subjective, although many connotative meanings of words are similar for certain cultures. The denotative meanings of words are usually based on facts, and the connotative meanings of words are usually based on emotion. Here are some examples of words and their denotative and connotative meanings in Western culture:

Word	Denotative Meaning	Connotative Meaning
Home	A permanent place where one lives, usually as a member of a family.	A place of warmth; a place of familiarity; comforting; a place of safety and security. "Home" usually has a positive connotation.
Snake	A long reptile with no limbs and strong jaws that moves along the ground; some snakes have a poisonous bite.	An evil omen; a slithery creature (human or nonhuman) that is deceitful or unwelcome. "Snake" usually has a negative connotation.
Winter	A season of the year that is the coldest, usually from December to February in the northern hemisphere and from June to August in the southern hemisphere.	Circle of life, especially that of death and dying; cold or icy; dark and gloomy; hibernation, sleep, or rest. Winter can have a negative connotation, although many who have access to heat may enjoy the snowy season from their homes.

Practice Questions

Synonyms

1. CONTRITE
    a. Tidy
    b. Unrealistic
    c. Remorseful
    d. Corrupt

2. ASSUAGE
    a. Irritate
    b. Persuade
    c. Soothe
    d. Redirect

3. TACIT
    a. Unspoken
    b. Shortened
    c. Tenuous
    d. Regal

Sentence Completion

4. The woman could not _____ the meaning of the entire conversation from the few words she overheard.
    a. Insult
    b. Avoid
    c. Decipher
    d. Measure

5. The writer _____ a story that detailed all of the happenings at summer camp.
    a. Discarded
    b. Adjourned
    c. Misaligned
    d. Composed

Answer Explanations

1. C: *Contrite* means to feel or express remorse, or to be regretful and interested in repenting. The noun *contrition* refers to severe remorse or penitence.

2. C: *Assuage* most nearly means to soothe or comfort, as in to assuage one's fears. It can also mean to lessen or make less severe, or to relieve. For example, an ice pack on a swollen knee may assuage the pain.

3. A: Something that is *tacit* is usually unspoken but implied. Tacit approval, for example, occurs when agreement or approval is understood without explicitly stating it.

4. C: Decipher. Choice *A*, *insult*, doesn't make sense within the context of the sentence. Choice *B*, *avoid*, is incorrect because the listener would not be trying to avoid the conversation while simultaneously trying to listen to what is being said. Choice *D*, *measure*, is not the best answer. Choice *C* is the correct answer because one deciphers or determines what was being said.

5. D: Composed. Choices *A* and *B* are incorrect because both are related to ending something or throwing something away. Choice *C* does not make sense within the context of the sentence. Choice *D* is correct because *compose* means "to create or write."

Quantitative Reasoning

Numbers and Operations

Base 10 System and Place Value

Our number system is known as a **place value system** because the placement of a digit within a number determines its value. For example, the number 345 has three values in three different places. The 3 is in the hundreds place, the 4 is in the tens place, and the 5 is in the ones place. After the hundreds place, each set of three values corresponds to the thousands, millions, billions, and trillions. For instance, consider the following whole number: 1,234,567,890. The 1 is in the billions place, the 2 is in the hundred-millions place, the 3 is in the ten-millions place, the 4 is in the millions place, the 5 is in the hundred-thousands place, the 6 is in the ten-thousands place, and the 7 is in the thousands place. This number is read "one billion, two hundred thirty-four million, five hundred sixty-seven thousand, eight hundred ninety." Notice that the commas are in the same place as in the number and that there is no "and" included anywhere.

If a number has parts of a whole, otherwise known as a fractional part, those parts can be represented with a decimal. The decimal point separates the whole number part from the fractional part. The first-place value to the right of the decimal is the tenths place, and the hundredths, thousandths, ten-thousandths, and hundred-thousandths places follow. Notice that there is no "oneths" place. Consider the decimal 4.567. The way to read a decimal is to read the whole number part, read the decimal as the word "and," read the part of the number to the right of the decimal as a normal whole number, and then finish with the place value name of the digit furthest to the right. For instance, 4.567 would be read "four and five hundred sixty-seven thousandths."

Here is a place value chart that shows the billions to the hundred-thousandths place values:

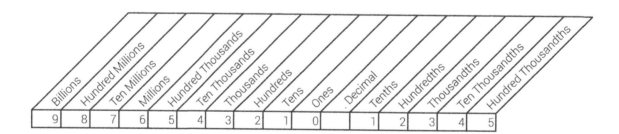

Identifying the Place and Value of a Digit

Each number in the base-10 system is made of the numbers 0—9, located in different places relative to the decimal point. Based on where the numbers fall, the value of a digit changes. For example, the number 7,509 has a seven in the thousands place. This means there are seven groups of one thousand. The number 457 has a seven in the ones place. This means there are seven groups of one. Even though there is a seven in both numbers, the place of the seven tells the value of the digit. A practice question may ask the place and value of the 4 in 3,948. The four is found in the tens place, which means four represents the number 40, or four groups of ten. Another place value may be on the opposite side of the decimal point. A question may ask the place and value of the 8 in the number 203.80. In this case, the eight is in the tenths place because it is in the first place to the right of the decimal point. It holds a value of eight-tenths, or eight groups of one-tenth.

Recognizing Relative Value of a Digit Given Its Place

The value of a digit is found by recognizing its place relative to the rest of the number. For example, the number 569.23 contains a 6. The position of the 6 is two places to the left of the decimal, putting it in the tens place. The tens place gives it a value of 60, or six groups of ten. The number 39.674 has a 4 in it. The number 4 is located three places to the right of the decimal point, placing it in the thousandths place. The value of the 4 is four-thousandths, because of its position relative to the other numbers and to the decimal. It can be described as 0.004 by itself, or four groups of one-thousandths. The numbers 100 and 0.1 are both made up of ones and zeros. The first number, 100, has a 1 in the hundreds place, giving it a value of one hundred. The second number, 0.1, has a 1 in the tenths place, giving that 1 a value of one-tenth. The place of the number gives it the value.

Basic Concepts of Number Theory

Whole numbers are the numbers 0, 1, 2, 3, ... Examples of other whole numbers would be 413 and 8,431. Notice that numbers such as 4.13 and $\frac{1}{4}$ are not included in whole numbers. **Counting numbers**, also known as **natural numbers**, consist of all whole numbers except for the zero. In set notation, the natural numbers are the set $\{1, 2, 3, ... \}$. The entire set of whole numbers and negative versions of those same numbers comprise the set of numbers known as **integers**. Therefore, in set notation, the integers are $\{..., -3, -2, -1, 0, 1, 2, 3, ... \}$. Examples of other integers are $-4,981$ and $90,131$. A number line is a great way to visualize the integers. Integers are labeled on the following number line:

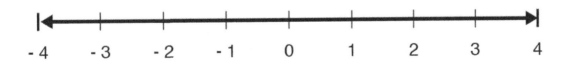

The arrows on the right- and left-hand sides of the number line show that the line continues indefinitely in both directions.

Fractions also exist on the number line as parts of a whole. For example, if an entire pie is cut into two pieces, each piece is half of the pie, or $\frac{1}{2}$. The top number in any fraction, known as the **numerator,** defines how many parts there are. The bottom number, known as the **denominator,** states how many pieces the whole is divided into. Fractions can also be negative or written in their corresponding decimal form.

A **decimal** is a number that uses a decimal point and numbers to the right of the decimal point representing the part of the number that is less than 1. For example, 3.5 is a decimal and is equivalent to the fraction $\frac{7}{2}$ or the mixed number $3\frac{1}{2}$. The decimal is found by dividing 2 into 7. Other examples of fractions are $\frac{2}{7}, \frac{-3}{14}$, and $\frac{14}{27}$.

Any number that can be expressed as a fraction is known as a **rational number**. Basically, if a and b are any integers and $b \neq 0$, then $\frac{a}{b}$ is a rational number. Any integer can be written as a fraction where the denominator is 1, so therefore the rational numbers consist of all fractions and all integers.

15

Any number that is not rational is known as an **irrational number.** Consider the number $\pi = 3.141592654\ldots$ The decimal portion of that number extends indefinitely. In that situation, a number can never be written as a fraction. Another example of an irrational number is $\sqrt{2} = 1.414213662\ldots$ Again, this number cannot be written as a ratio of two integers.

Together, the set of all rational and irrational numbers makes up the real numbers. The number line contains all real numbers. To graph a number other than an integer on a number line, it needs to be plotted between two integers. For example, 3.5 would be plotted halfway between 3 and 4.

Even numbers are integers that are divisible by 2. For example, 6, 100, 0, and -200 are all even numbers. Odd numbers are integers that are not divisible by 2. If an odd number is divided by 2, the result is a fraction. For example, -5, 11, and -121 are odd numbers.

Prime numbers consist of natural numbers greater than 1 that are not divisible by any other natural numbers other than themselves and 1. For example, 3, 5, and 7 are prime numbers. If a natural number is not prime, it is known as a **composite number**. 8 is a composite number because it is divisible by both 2 and 4, which are natural numbers other than itself and 1.

The **absolute value** of any real number is the distance from that number to 0 on the number line. The absolute value of a number can never be negative. For example, the absolute value of both 8 and -8 is 8 because they are both 8 units away from 0 on the number line. This is written as $|8| = |-8| = 8$.

Ordering Real Numbers

Ordering rational numbers is a way to compare two or more different numerical values. Determining whether two amounts are equal, less than, or greater than is the basis for comparing both positive and negative numbers. Also, a group of numbers can be compared by ordering them from the smallest amount to the largest amount. A few symbols are necessary to use when ordering rational numbers. The equals sign, $=$, shows that the two quantities on either side of the symbol have the same value. For example, $\frac{12}{3} = 4$ because both values are equivalent. Another symbol that is used to compare numbers is $<$, which represents "less than." With this symbol, the smaller number is placed on the left and the larger number is placed on the right. Always remember that the symbol's "mouth" opens up to the larger number. When comparing negative and positive numbers, it is important to remember that the number occurring to the left on the number line is always smaller and is placed to the left of the symbol. This idea might seem confusing because some values could appear at first glance to be larger, even though they are not. For example, $-5 < 4$ is read "negative 5 is less than 4."

Here is an image of a number line for help:

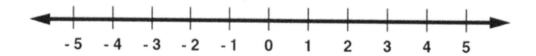

16

The symbol $\leq$ represents "less than or equal to," and it joins $<$ with equality. Therefore, both $-5 \leq 4$ and $-5 \leq -5$ are true statements and "-5 is less than or equal to both 4 and -5." Other symbols are $>$ and $\geq$, which represent "greater than" and "greater than or equal to." Both $4 \geq -1$ and $-1 \geq -1$ are correct ways to use these symbols.

Here is a table of these four inequality symbols:

Symbol		Definition
	$<$	less than
	$\leq$	less than or equal to
	$>$	greater than
	$\geq$	greater than or equal to

Comparing integers is a straightforward process, especially when using the number line, but the comparison of decimals and fractions is not as obvious. When comparing two non-negative decimals, compare digit by digit, starting from the left. The larger value contains the first larger digit. For example, 0.1456 is larger than 0.1234 because the value of the number in the hundredths place (4, in this case) in the first decimal is larger than the value of the number in the hundredths place (2) in the second decimal. When comparing a fraction with a decimal, convert the fraction to a decimal and then compare in the same manner. Finally, there are a few options when comparing fractions.

If two non-negative fractions have the same denominator, the fraction with the larger numerator is the larger value. If they have different denominators, they can be converted to equivalent fractions with a common denominator to be compared, or they can be converted to decimals to be compared. When comparing two negative decimals or fractions, a different approach must be used. It is important to remember that the smaller number exists to the left on the number line. Therefore, when comparing two negative decimals by place value, the number with the larger first place value is smaller due to the negative sign. Whichever value is closer to 0 is larger. For example, -0.456 is larger than -0.498 because of the values in the hundredth places. If two negative fractions have the same denominator, the fraction with the larger numerator is smaller because of the negative sign.

Placing Numbers on a Number Line

A **number line** is a tool used to compare numbers by showing where they fall in relation to one another. Labeling a number line with integers is simple because they have no fractional component and the values are easier to understand. The number line may start at -3 and go up to -2, then -1, then 0, and 1, 2, 3. This order shows that number 2 is larger than -1 because it falls further to the right on the number line. When positioning rational numbers, the process may take more time because it requires that they all be in the same form. If they are percentages, fractions, and decimals, then conversions will have to be made to put them in the same form.

For example, if the numbers $\frac{5}{4}$, 45%, and 2.38 need to be put in order on a number line, the numbers must first be transformed into one single form. Decimal form is an easy common ground because fractions can be changed by simply dividing and percentages can be changed by moving the decimal point. After conversions are made, the list becomes 1.25, 0.45, and 2.38 respectively. Now the list is easier to arrange. The number line with the list in order is shown in the top half of the graphic below in the order 0.45, 1.25, and 2.38.

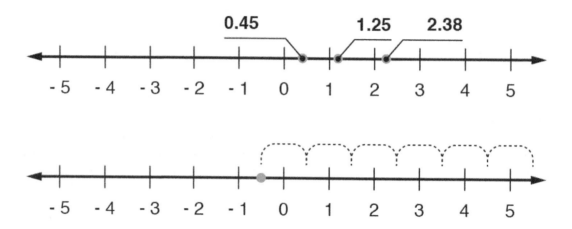

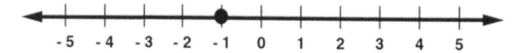

0 is the dividing point between the positive and negative numbers on a number line. In order to plot numbers on a number line, the number is first located on the line, and then a dot is drawn at the location of the number. For example, here is -1 graphed on a number line:

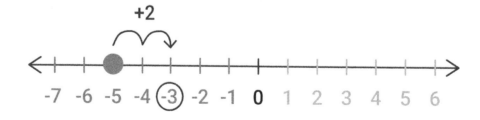

The number line can be used to add and subtract numbers. In order to add a positive number to a positive number, locate the first number on the number line and then move to the right the number of units corresponding to the second number. For example, $-5 + 2 = -3$. First, -5 is located on the number line. Then, moving 2 units to the right results in a value of -3.

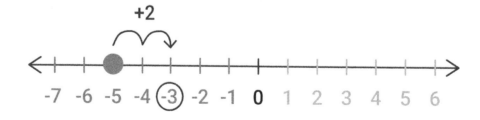

In order to add a negative number to a number, plot the first point on the number line and then move to the left the number of units corresponding to the absolute value of the second number. For example, 8 +

$(-2) = 6$. First, 8 is located on the number line. Then, moving |-2|=2 units to the left results in a value of 6.

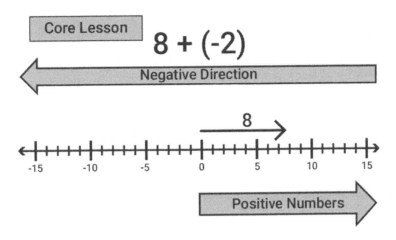

Because subtracting a positive number is the same as adding a negative number, this same procedure could be used to obtain $8 - 2 = 6$. Subtracting a positive number also involves moving that many units to the left on the number line.

Composing and Decomposing Multi-Digit Numbers

Composing and decomposing numbers reveals the place value held by each number 0 through 9 in each position. For example, the number 17 is read as "seventeen." It can be decomposed into the numbers 10 and 7. It can be described as 1 group of ten and 7 ones. The one in the tens place represents one set of ten. The seven in the ones place represents seven sets of one. Added together, they make a total of seventeen. The number 48 can be written in words as "forty-eight." It can be decomposed into the numbers 40 and 8, where there are 4 groups of ten and 8 groups of one. The number 296 can be decomposed into 2 groups of one hundred, 9 groups of ten, and 6 groups of one. There are two hundreds, nine tens, and six ones. Decomposing and composing numbers lays the foundation for visually picturing the number and its place value, and adding and subtracting multiple numbers with ease.

Rounding Multi-Digit Numbers

Numbers can be rounded by recognizing the place value where the rounding takes place, then looking at the number to the right. If the number to the right is five or greater, the number to be rounded goes up one. If the number to the right is four or less, the number to be rounded stays the same. For example, the number 438 can be rounded to the tens place. The number 3 is in the tens place and the number to the right is 8. Because the 8 is 5 or greater, the 3 then rounds up to a 4. The rounded number is 440. Another number, 1,394, can be rounded to the thousands place. The number in the thousands place is 1, and the number to the right is 3. As the 3 is 4 or less, it means the 1 stays the same and the rounded number is 1,000. Rounding is also a form of estimating. The number 9.58 can be rounded to the tenths place. The number 5 is in the tenths place, and the number 8 is to the right of it. Because 8 is 5 or greater, the 5 changes to a 6. The rounded number becomes 9.6.

Sometimes it is helpful to round answers that are in decimal form. First, find the place to which the rounding needs to be done. Then, look at the digit to the right of it. If that digit is 4 or less, the number in the place value to its left stays the same, and everything to its right becomes a 0. This process is known as rounding down. If that digit is 5 or higher, round up by increasing the place value to its left by 1, and every number to its right becomes a 0. If those 0's are in decimals, they can be dropped. For example, 0.145 rounded to the nearest hundredth place would be rounded up to 0.15, and 0.145 rounded to the nearest tenth place would be rounded down to 0.1.

Estimation

Sometimes it is helpful to find an estimated answer to a problem rather than working out an exact answer. An estimation might be much quicker to find, and given the scenario, an estimation might be all that is required. For example, if Aria goes grocery shopping and has only a $100 bill to cover all of her purchases, it might be appropriate for her to estimate the total of the items she is purchasing to determine if she has enough money to cover them. Also, an estimation can help determine if an answer makes sense. For instance, if an answer in the 100s is expected, but the result is a fraction less than 1, something is probably wrong in the calculation.

The first type of estimation involves rounding. As mentioned, **rounding** consists of expressing a number in terms of the nearest decimal place like the tenth, hundredth, or thousandth place, or in terms of the nearest whole number unit like tens, hundreds, or thousands place. When rounding to a specific place value, look at the digit to the right of the place. If it is 5 or higher, round the number to its left up to the next value, and if it is 4 or lower, keep that number at the same value. For instance, 1,654.2674 rounded to the nearest thousand is 2,000, and the same number rounded to the nearest thousandth is 1,654.267. Rounding can be used in the scenario when grocery totals need to be estimated. Items can be rounded to the nearest dollar. For example, a can of corn that costs $0.79 can be rounded to $1.00, and then all other items can be rounded in a similar manner and added together.

When working with larger numbers, it might make more sense to round to higher place values. For example, when estimating the total value of a dealership's car inventory, it would make sense to round the car values to the nearest thousands place. The price of a car that is on sale for $15,654 can be estimated at $16,000. All other cars on the lot could be rounded in the same manner, and then their sum can be found. Depending on the situation, it might make sense to calculate an over-estimate. For example, to make sure Aria has enough money at the grocery store, rounding up every time for each item would ensure that she will have enough money when it comes time to pay. A $0.40 item rounded up to $1.00 would ensure that there is a dollar to cover that item. Traditional rounding rules would round $0.40 to $0, which does not make sense in this particular real-world setting. Aria might not have a dollar available at checkout to pay for that item if she uses traditional rounding. It is up to the customer to decide the best approach when estimating.

Estimating is also very helpful when working with measurements. Bryan is updating his kitchen and wants to retile the floor. Again, an over-measurement might be useful. Also, rounding to nearest half-unit might be helpful. For instance, one side of the kitchen might have an exact measurement of 14.32 feet, and the most useful measurement needed to buy tile could be estimating this quantity to be 14.5 feet. If the kitchen was rectangular and the other side measured 10.9 feet, Bryan might round the other side to 11 feet. Therefore, Bryan would find the total tile necessary according to the following area calculation:

$$14.5 \times 11 = 159.5 \text{ square feet}$$

To make sure he purchases enough tile, Bryan would probably want to purchase at least 160 square feet of tile. This is a scenario in which an estimation might be more useful than an exact calculation. Having more tile than necessary is better than having an exact amount, in case any tiles are broken or otherwise unusable.

Finally, estimation is helpful when exact answers are necessary. Consider a situation in which Sabina has many operations to perform on numbers with decimals, and she is allowed a calculator to find the result. Even though an exact result can be obtained with a calculator, there is always a possibility that Sabina could make an error while inputting the data. For example, she could miss a decimal place, or misuse a parenthesis, causing a problem with the actual order of operations. In this case, a quick estimation at the beginning would be helpful to make sure the final answer is given with the correct number of units. Sabina has to find the exact total of 10 cars listed for sale at the dealership. Each price has two decimal places included to account for both dollars and cents. If one car is listed at $21,234.43 but Sabina incorrectly inputs into the calculator the price of $2,123.443, this error would throw off the final sum by almost $20,000. A quick estimation at the beginning, by rounding each price to the nearest thousands place and finding the sum of the prices, would give Sabina an amount to compare the exact amount to. This comparison would let Sabina see if an error was made in her exact calculation.

Factorization

Factorization is the process of breaking up a mathematical quantity, such as a number or polynomial, into a product of two or more factors. For example, a factorization of the number 16 is:

$$16 = 8 \times 2$$

If multiplied out, the factorization results in the original number. A **prime factorization** is a specific factorization when the number is factored completely using prime numbers only. For example, the prime factorization of 16 is:

$$16 = 2 \times 2 \times 2 \times 2$$

A factor tree can be used to find the prime factorization of any number. Within a factor tree, pairs of factors are found until no other factors can be used, as in the following factor tree of the number 84:

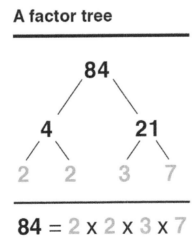

A factor tree

$$84 = 2 \times 2 \times 3 \times 7$$

It first breaks 84 into 21 × 4, which is not a prime factorization. Then, both 21 and 4 are factored into their primes. The final numbers on each branch consist of the numbers within the prime factorization. Therefore:

$$84 = 2 \times 2 \times 3 \times 7$$

Factorization can be helpful in finding greatest common divisors and least common denominators.

Also, a factorization of an algebraic expression can be found. Throughout the process, a more complicated expression can be decomposed into products of simpler expressions. To factor a polynomial, first determine if there is a greatest common factor. If there is, factor it out. For example, $2x^2 + 8x$ has a greatest common factor of $2x$ and can be written as $2x(x + 4)$.

Operations on Rational Numbers

The four basic operations include addition, subtraction, multiplication, and division. The result of addition is a **sum**, the result of subtraction is a **difference**, the result of multiplication is a **product**, and the result of division is a **quotient**. Each type of operation can be used when working with rational numbers; however, the basic operations need to be understood first while using simpler numbers before working with fractions and decimals.

Performing these operations should first be learned using whole numbers. Addition needs to be done column by column. To add two whole numbers, add the ones column first, then the tens columns, then the hundreds, etc. If the sum of any column is greater than 9, a one must be carried over to the next column. For example, the following is the result of $482 + 924$:

$$
\begin{array}{r}
1 \\
482 \\
+924 \\
\hline
1406
\end{array}
$$

Notice that the sum of the ten's column was 10, so a one was carried over to the hundred's column. Subtraction is also performed column by column. Subtraction is performed in the one's column first, then the tens, etc. If the number on top is less than the number below, a one must be borrowed from the column to the left. For example, the following is the result of $5,424 - 756$:

$$
\begin{array}{r}
4\ \ 13\ \ 11\ \ 14 \\
\cancel{5}\ \ \cancel{4}\ \ \cancel{2}\ \ 4 \\
-\ \ 7\ \ \ 5\ \ \ 6 \\
\hline
4\ \ \ 6\ \ \ 6\ \ \ 8
\end{array}
$$

Notice that a one is borrowed from the tens, hundreds, and thousands place. After subtraction, the answer can be checked through addition. A check of this problem would be to show that $756 + 4,668 = 5,424$.

Multiplication of two whole numbers is performed by writing one on top of the other. The number on top is known as the **multiplicand,** and the number below is the **multiplier**. Perform the multiplication by multiplying the multiplicand by each digit of the multiplier. Make sure to place the ones value of each result under the multiplying digit in the multiplier. Each value to the right is then a 0. The product is found by adding each product. For example, the following is the process of multiplying 46 times 37 where 46 is the multiplicand and 37 is the multiplier:

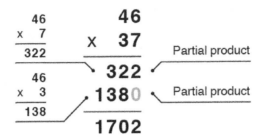

Finally, division can be performed using long division. When dividing a number by another number, the first number is known as the **dividend,** and the second is the **divisor.** For example, with $a \div b = c$, a is the dividend, b is the divisor, and c is the quotient. For long division, place the dividend within the division symbol and the divisor on the outside. For example, with $8,764 \div 4$, refer to the first problem in the diagram below. First, there are two 4's in the first digit, 8. This number 2 gets written above the 8. Then, multiply 4 times 2 to get 8, and that product goes below the 8. Subtract to get 8, and then carry down the 7. Continue the same steps. $7 \div 4 = 1$ R3, so 1 is written above the 7. Multiply 4 times 1 to get 4, and write it below the 7. Subtract to get 3, and carry the 6 down next to the 3. Resulting steps give a 9 and a 1. The final subtraction results in a 0, which means that 8,764 is divisible by 4. There are no remaining numbers.

The second example shows that:

$$4,536 \div 216 = 21$$

The steps are a little different because 216 cannot be contained in 4 or 5, so the first step is placing a 2 above the 3 because there are two 216's in 453. Finally, the third example shows that:

$$546 \div 31 = 17 \, R19$$

The 19 is a remainder. Notice that the final subtraction does not result in a 0, which means that 546 is not divisible by 31.

The remainder can also be written as a fraction over the divisor to say that:

$$546 \div 31 = 17\frac{19}{31}$$

```
   2191            21              17 r 19
 4│8764        216│4536         31│546
   8↓              432↓             31↓
   07             216             236
    4↓            216             217
    36              0              19
    36
    04
     4
     0
```

If a division problem relates to a real-world application, and a remainder does exist, it can have meaning. For example, consider the third example:

$$546 \div 31 = 17\,R19$$

Let's say that we had $546 to spend on calculators that cost $31 each, and we wanted to know how many we could buy. The division problem would answer this question. The result states that 17 calculators could be purchased, with $19 left over. Notice that the remainder will never be greater than or equal to the divisor.

Once the operations are understood with whole numbers, they can be used with integers. There are many rules surrounding operations with negative numbers. First, consider addition with integers. The sum of two numbers can first be shown using a number line. For example, to add $-5 + (-6)$, plot the point -5 on the number line. Then, because a negative number is being added, move 6 units to the left. This process results in landing on -11 on the number line, which is the sum of -5 and -6. If adding a positive number, move to the right. Visualizing this process using a number line is useful for understanding; however, it is not efficient. A quicker process is to learn the rules. When adding two numbers with the same sign, add the absolute values of both numbers, and use the common sign of both numbers as the sign of the sum. For example, to add $-5 + (-6)$, add their absolute values:

$$5 + 6 = 11$$

Then, introduce a negative number because both addends are negative. The result is -11. To add two integers with unlike signs, subtract the lesser absolute value from the greater absolute value, and apply the sign of the number with the greater absolute value to the result. For example, the sum $-7 + 4$ can be computed by finding the difference $7 - 4 = 3$ and then applying a negative because the value with the larger absolute value is negative. The result is -3. Similarly, the sum $-4 + 7$ can be found by computing the same difference but leaving it as a positive result because the addend with the larger absolute value is positive. Also, recall that any number plus 0 equals that number. This is known as the **Addition Property of 0.**

Subtracting two integers can be computed by changing to addition to avoid confusion. The rule is to add the first number to the opposite of the second number. The opposite of a number is the number on the other side of 0 on the number line, which is the same number of units away from 0. For example, −2 and 2 are opposites. Consider $4 - 8$. Change this to adding the opposite as follows: $4 + (-8)$. Then, follow the rules of addition of integers to obtain −4. Secondly, consider $-8 - (-2)$. Change this problem to adding the opposite as $-8 + 2$, which equals −6. Notice that subtracting a negative number functions the same as adding a positive number.

Multiplication and division of integers are actually less confusing than addition and subtraction because the rules are simpler to understand. If two factors in a multiplication problem have the same sign, the result is positive. If one factor is positive and one factor is negative, the result, known as the **product**, is negative. For example, $(-9)(-3) = 27$ and:

$$9(-3) = -27$$

Also, any number times 0 always results in 0. If a problem consists of more than a single multiplication, the result is negative if it contains an odd number of negative factors, and the result is positive if it contains an even number of negative factors. For example:

$$(-1)(-1)(-1)(-1) = 1 \text{ and } (-1)(-1)(-1)(-1)(-1) = 1$$

These two examples of multiplication also bring up another concept. Both are examples of repeated multiplication, which can be written in a more compact notation using exponents.

The first example can be written as $(-1)^4 = 1$, and the second example can be written as:

$$(-1)^5 = -1$$

Both are exponential expressions, −1 is the base in both instances, and 4 and 5 are the respective exponents. Note that a negative number raised to an odd power is always negative, and a negative number raised to an even power is always positive. Also, $(-1)^4$ is not the same as -1^4. In the first expression, the negative is included in the parentheses, but it is not in the second expression. The second expression is found by evaluating 1^4 first to get 1 and then by applying the negative sign to obtain −1.

A similar theory applies within division. First, consider some vocabulary. When dividing 14 by 2, it can be written in the following ways:

$$14 \div 2 = 7 \text{ or } \frac{14}{2} = 7$$

14 is the **dividend,** 2 is the **divisor,** and 7 is the **quotient.** If two numbers in a division problem have the same sign, the quotient is positive. If two numbers in a division problem have different signs, the quotient is negative. For example, $14 \div (-2) = -7$, and $-14 \div (-2) = 7$. To check division, multiply the quotient by the divisor to obtain the dividend. Also, remember that 0 divided by any number is equal to 0. However, any number divided by 0 is undefined. It just does not make sense to divide a number by 0 parts.

Once the rules for integers are understood, move on to learning how to perform operations with fractions and decimals. Recall that a rational number can be written as a fraction and can be converted to a decimal through division. If a rational number is negative, the rules for adding, subtracting, multiplying, and dividing integers must be used. If a rational number is in fraction form, performing addition, subtraction, multiplication, and division is more complicated than when working with integers. First, consider addition. To add two fractions having the same denominator, add the numerators and then reduce the fraction. When an answer is a fraction, it should always be in lowest terms. **Lowest terms** means that every common factor, other than 1, between the numerator and denominator is divided out.

For example:

$$\frac{2}{8} + \frac{4}{8} = \frac{6}{8} = \frac{6 \div 2}{8 \div 2} = \frac{3}{4}$$

Both the numerator and denominator of $\frac{6}{8}$ have a common factor of 2, so 2 is divided out of each number to put the fraction in lowest terms. If denominators are different in an addition problem, the fractions must be converted to have common denominators. The **least common denominator (LCD)** of all the given denominators must be found, and this value is equal to the **least common multiple (LCM)** of the denominators. This non-zero value is the smallest number that is a multiple of both denominators. Then, rewrite each original fraction as an equivalent fraction using the new denominator. Once in this form, apply the process of adding with like denominators. For example, consider $\frac{1}{3} + \frac{4}{9}$. The LCD is 9 because it is the smallest multiple of both 3 and 9. The fraction $\frac{1}{3}$ must be rewritten with 9 as its denominator. Therefore, multiply both the numerator and denominator by 3. Multiplying by $\frac{3}{3}$ is the same as multiplying by 1, which does not change the value of the fraction. Therefore, an equivalent fraction is $\frac{3}{9}$, and $\frac{1}{3} + \frac{4}{9} = \frac{3}{9} + \frac{4}{9} = \frac{7}{9}$, which is in lowest terms. Subtraction is performed in a similar manner; once the denominators are equal, the numerators are then subtracted.

The following is an example of addition of a positive and a negative fraction:

$$-\frac{5}{12} + \frac{5}{9}$$

$$-\frac{5 \times 3}{12 \times 3} + \frac{5 \times 4}{9 \times 4}$$

$$-\frac{15}{36} + \frac{20}{36} = \frac{5}{36}$$

Common denominators are not used in multiplication and division. To multiply two fractions, multiply the numerators together and the denominators together. Then, write the result in lowest terms. For example:

$$\frac{2}{3} \times \frac{9}{4} = \frac{18}{12} = \frac{3}{2}$$

Alternatively, the fractions could be factored first to cancel out any common factors before performing the multiplication. For example:

$$\frac{2}{3} \times \frac{9}{4} = \frac{2}{3} \times \frac{3 \times 3}{2 \times 2} = \frac{3}{2}$$

This second approach is helpful when working with larger numbers, as common factors might not be obvious. Multiplication and division of fractions are related because the division of two fractions is changed into a multiplication problem. This means that dividing a fraction by another fraction is the same as multiplying the first fraction by the reciprocal of the second fraction, so that second fraction must be inverted, or "flipped," to be in reciprocal form. For example:

$$\frac{11}{15} \div \frac{3}{5} = \frac{11}{15} \times \frac{5}{3}$$

$$\frac{55}{45} = \frac{11}{9}$$

The fraction $\frac{5}{3}$ is the reciprocal of $\frac{3}{5}$.

It is possible to multiply and divide numbers containing a mix of integers and fractions. In this case, convert the integer to a fraction by placing it over a denominator of 1. For example, a division problem involving an integer and a fraction is

$$3 \div \frac{1}{2} = \frac{3}{1} \times \frac{2}{1} = \frac{6}{1} = 6$$

Finally, when performing operations with rational numbers that are negative, the same rules apply as when performing operations with integers. For example, a negative fraction multiplied by a negative fraction results in a positive value, and a negative fraction subtracted from a negative fraction results in a negative value.

Operations can be performed on rational numbers in decimal form. Recall that to write a fraction as an equivalent decimal expression, divide the numerator by the denominator. For example:

$$\frac{1}{8} = 1 \div 8 = 0.125$$

With the case of decimals, it is important to keep track of place value. To add decimals, make sure the decimal places are in alignment so that the numbers are lined up with their decimal points and add vertically. If the numbers do not line up because there are extra or missing place values in one of the numbers, then zeros may be used as placeholders. For example, $0.123 + 0.23$ becomes:

$$
\begin{array}{r}
0.123 \\
+\ 0.230 \\
\hline
0.353
\end{array}
$$

Subtraction is done the same way. Multiplication and division are more complicated. To multiply two decimals, place one on top of the other as in a regular multiplication process and do not worry about lining up the decimal points. Then, multiply as with whole numbers, ignoring the decimals. Finally, in the solution, insert the decimal point as many places to the left as there are total decimal values in the original problem. Here is an example of a decimal multiplication problem:

$$
\begin{array}{rl}
0.52 & \textit{2 decimal places} \\
\times\ \ 0.2 & \textit{1 decimal place} \\
\hline
0.104 & \textit{3 decimal places}
\end{array}
$$

The answer to 52 times 2 is 104, and because there are three decimal values in the problem, the decimal point is positioned three units to the left in the answer.

The decimal point plays an integral role throughout the whole problem when dividing with decimals. First, set up the problem in a long division format. If the divisor is not an integer, the decimal must be moved to the right as many units as needed to make it an integer. The decimal in the dividend must be moved to the right the same number of places to maintain equality. Then, division is completed normally.

Here is an example of long division with decimals:

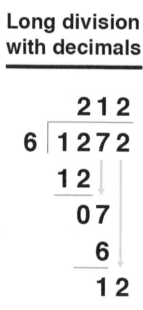

Long division with decimals

```
       2 1 2
   ┌─────────
 6 │ 1 2 7 2
   │ 1 2
   │ ───
   │   0 7
   │     6
   │     ───
   │     1 2
```

Because the decimal point is moved two units to the right in the divisor of 0.06 to turn it into the integer 6, it is also moved two units to the right in the dividend of 12.72 to make it 1,272. The result is 212, and remember that a division problem can always be checked by multiplying the answer by the divisor to see if the result is equal to the dividend.

Interpreting Remainders in Division Problems

Understanding remainders begins with understanding the division problem. The problem $24 \div 7$ can be read as "twenty-four divided by seven." The problem is asking how many groups of 7 will fit into 24. Counting by seven, the multiples are 7, 14, 21, 28. Twenty-one, which is three groups of 7, is the closest to 24. The difference between 21 and 24 is 3, which is called the remainder. This is a remainder because it is the number that is left out after the three groups of seven are taken from 24. The answer to this division problem can be written as 3 with a remainder 3, or $3\frac{3}{7}$. The fraction $\frac{3}{7}$ can be used because it shows the part of the whole left when the division is complete. Another division problem may have the following numbers: $36 \div 5$. This problem is asking how many groups of 5 will fit evenly into 36. When counting by multiples of 5, the following list is generated: 5, 10, 15, 20, 25, 30, 35, 40. As seen in the list, there are seven groups of five that make 35. To get to the total of 36, there needs to be one additional number. The answer to the division problem would be $36 \div 5 = 7$ R1, or $7\frac{1}{5}$. The fractional part represents the number that cannot make up a whole group of five.

Order of Operations

If more than one operation is to be completed in a problem, follow the Order of Operations. The mnemonic device, PEMDAS, for the order of operations states the order in which addition, subtraction, multiplication, and division needs to be done. It also includes when to evaluate operations within grouping symbols and when to incorporate exponents. PEMDAS, which some remember by thinking "please excuse my dear Aunt Sally," refers to parentheses, exponents, multiplication, division, addition, and subtraction. First, within an expression, complete any operation that is within parentheses, or any other grouping symbol like brackets, braces, or absolute value symbols. Note that this does not refer to the case when parentheses are used to represent multiplication like $(2)(5)$. An operation is not within parentheses like it is in (2×5). Then, any exponents must be computed. Next, multiplication and division are performed from left to right.

Finally, addition and subtraction are performed from left to right. The following is an example in which the operations within the parentheses need to be performed first, so the order of operations must be applied to the exponent, subtraction, addition, and multiplication within the grouping symbol:

$$9 - 3(3^2 - 3 + 4 \cdot 3)$$

$$9 - 3(3^2 - 3 + 4 \cdot 3) \quad \text{Work within the parentheses first}$$

$$= 9 - 3(9 - 3 + 12)$$

$$= 9 - 3(18)$$

$$= 9 - 54$$

$$= -45$$

Properties of Operations

Operations follow certain properties and rules. Addition and multiplication follow the **commutative property.** This means that the numbers can be added or multiplied in any order, and the result will be the same. For example, $6 \times 3 = 18$ and $3 \times 6 = 18$. Subtraction and division are not commutative. Addition and multiplication are also associative. The **associative property** means that the grouping of numbers with parentheses does not change the answer to the problem. For example, $(7 + 8) + 5 = 20$ and $(7 + 5) + 8 = 20$. Another property that impacts operations is the distributive property. The **distributive property** states that when a sum or difference inside the parentheses is multiplied by a number, it is the same as multiplying both the numbers inside the parentheses by the outside number and adding or subtracting the results. For example:

$$8 \times (4 + 3) = 8 \times 7 = 56 \text{ and } 8 \times 4 + 8 \times 3 = 32 + 24 = 56$$

Rational Numbers and Their Operations

Rational numbers can be whole or negative numbers, fractions, or repeating decimals because these numbers can all be written as fractions. Whole numbers can be written as fractions; for example, 25 and 17 can be written as $\frac{25}{1}$ and $\frac{17}{1}$. One way of interpreting these fractions is to say that they are **ratios**, or comparisons of two quantities. The fractions given may represent 25 students to 1 classroom, or 17 desks to 1 computer lab. Repeating decimals can also be written as fractions of integers, such as 0.3333 and 0.6666667. These repeating decimals can be written as the fractions $\frac{1}{3}$ and $\frac{2}{3}$. Fractions can be described as having a part-to-whole relationship. The fraction $\frac{1}{3}$ may represent 1 piece of pizza out of the whole cut into 3 pieces. The fraction $\frac{2}{3}$ may represent 2 pieces of the same whole pizza. Adding the fractions $\frac{1}{3}$ and $\frac{2}{3}$ is as simple as adding the numerators, 1 and 2, because the denominator on both fractions is 3. This means the numbers in the numerators are referring to multiples of the same size piece of pizza. When adding these fractions, the result is $\frac{3}{3}$, or 1. Both of these numbers are rational and represent a whole, or in this problem, a whole pizza.

Other than fractions, rational numbers also include whole numbers and negative integers. When whole numbers are added, other than zero, the result is always greater than the addends. For example, the equation $4 + 18 = 22$ shows 4 increased by 18, with a result of 22. When subtracting rational numbers, sometimes the result is a negative number. For example, the equation $5 - 12 = -7$ shows that taking 12 away from 5 results in a negative answer because 5 is smaller than 12. The difference is -7 because the starting number is smaller than the number taken away. For multiplication and division, similar results are found. Multiplying rational numbers may look like the following equation: $5 \times 7 = 35$, where both numbers are positive and whole, and the result is a larger number than the factors. The number 5 is counted 7 times, which results in a total of 35. Sometimes, the equation looks like $-4 \times 3 = -12$, so the result is negative because a positive number times a negative number gives a negative answer. The rule is that any time a negative number and a positive number are multiplied or divided, the result is negative.

Representing Rational Numbers and Their Operations

Concrete models create a way of thinking about math that generates learning on a more permanent level. Memorizing abstract math formulas will not create a lasting effect on the brain. The following picture shows fractions represented by Lego blocks. By starting with the whole block of eight, it can be split into half, which is a four-block, and a fourth, which is a two-block. The one-eighth representation is a single block.

After splitting these up, addition and subtraction can be performed by adding or taking away parts of the blocks. Different combinations of fractions can be used to make a whole, or taken away to make various parts of a whole.

Using Colored Blocks to Model Functions

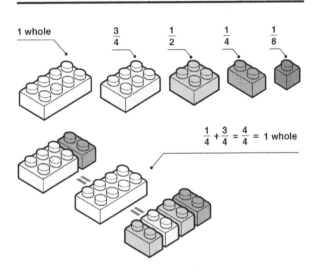

Multiplication can also be done using array models. The following picture shows a model of multiplying 3 times 4. **Arrays** are formed when the first factor is shown in a row. The second factor is shown in that number of columns. When the rectangle is formed, the blocks fill in to make a total, or the result of multiplication. The three rows and four columns show each factor and when the blocks are filled in, the total is twelve. Arrays are great ways to represent multiplication because they show each factor, and where the total comes from, with rows and columns.

Multiplication Array Model for 3 x 4

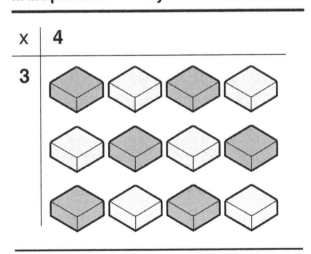

3 x 4 = 12

Another representation of fractions is shown below in the pie charts. Moving from whole numbers to part of numbers with fractions can be a concept that is difficult to grasp. Starting with a whole pie and splitting it into parts can be helpful with generating fractions. The first pie shows quarters or sections that are one-fourth because it is split into four parts. The second pie shows parts that equal one-fifth because it is split into five parts. Pies can also be used to add fractions. If $\frac{1}{5}$ and $\frac{1}{4}$ are being added, a common denominator must be found by splitting the pies into the same number of parts. The same number of parts can be found by determining the least common multiple. For 4 and 5, the least common multiple is 20. The pies can be split until there are 20 parts. The same portion of the pie can be shaded for each fraction and then added together to find the sum.

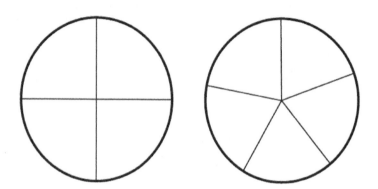

Multiplication and division can be represented by equations. These equations show the numbers involved and the operation. For example, "eight multiplied by six equals forty-eight" is seen in the following equation: $8 \times 6 = 48$. This operation can be modeled by rectangular arrays where one factor, 8, is the number of rows, and the other factor, 6, is the number of columns, as follows:

Array of 8 x 6 = 48

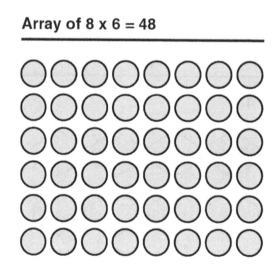

Rectangular arrays show what happens with the concept of multiplication. As one row of dots is drawn, that represents the first factor in the problem. Then the second factor is used to add the number of columns. The final model includes six rows of eight columns which results in forty-eight dots. These rectangular arrays show how multiplication of whole numbers will result in a number larger than the factors.

Division can also be represented by equations and area models. A division problem such as "twenty-four divided by three equals eight" can be written as the following equation: $24 \div 8 = 3$. The object below shows an area model to represent the equation. As seen in the model, the whole box represents 24 and the 3 sections represent the division by 3. In more detail, there could be 24 dots written in the whole box and each box could have 8 dots in it. Division shows how numbers can be divided into groups. For the example problem, it is asking how many numbers will be in each of the 3 groups that together make 24. The answer is 8 in each group.

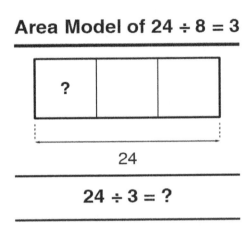

Area Model of 24 ÷ 8 = 3

24

$24 \div 3 = ?$

Using Whole-Number Exponents to Denote Powers of 10

Numbers can also be written using exponents. The number 7,000 can be written as $7 \times 1,000$ because 7 is in the thousands place. It can also be written as 7×10^3 because $1,000 = 10^3$. Another number that can use this notation is 500. It can be written as 5×100, or 5×10^2, because $100 = 10^2$. The number 30 can be written as 3×10, or 3×10^1, because $10 = 10^1$. Notice that each one of the exponents of 10 is equal to the number of zeros in the number. Seven is in the thousands place, with three zeros, and the exponent on ten is 3. The five is in the hundreds place, with two zeros, and the exponent on the ten is 2. A question may give the number 40,000 and ask for it to be rewritten using exponents with a base of ten. Because the number has a four in the ten-thousands place and four zeros, it can be written using an exponent of four: 4×10^4.

Composing and Decomposing Fractions

Fractions are ratios of whole numbers and their negatives. Fractions represent parts of wholes, whether pies, or money, or work. The number on top, or numerator, represents the part, and the bottom number, or denominator, represents the whole. The number $\frac{1}{2}$ represents half of a whole. Other ways to represent one-half are $\frac{2}{4}, \frac{3}{6}$, and $\frac{5}{10}$. These are fractions not written in simplest form, but the numerators are all halves of the denominators. The fraction $\frac{1}{4}$ represents 1 part to a whole of 4 parts. This can be modeled by the quarter's value in relation to the dollar. One quarter is $\frac{1}{4}$ of a dollar. In the same way, 2 quarters make up $\frac{1}{2}$ of a dollar, so 2 fractions of $\frac{1}{4}$ make up a fraction of $\frac{1}{2}$. Three quarters make up three-fourths of a dollar. The three fractions of $\frac{1}{4} + \frac{1}{4} + \frac{1}{4}$ are equal to $\frac{3}{4}$ of a whole. This illustration can be seen using the bars below divided into one whole, then two halves, then three sections of one-third, then four sections of

33

one-fourth. Based on the size of the fraction, different numbers of each fraction are needed to make up a whole.

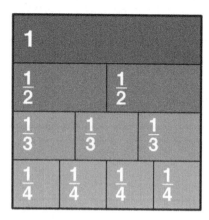

Unit Fractions

A **unit fraction** is a fraction where the numerator has a value of one. The fractions one-half, one-third, one-seventh, and one-tenth are all examples of unit fractions. Nonexamples of unit fractions include three-fourths, four-fifths, and seven-twelfths. The value of unit fractions changes as the denominator changes, because the numerator is always one. The unit fraction one-half requires two parts to make a whole. The unit fraction one-third requires three parts to make a whole. In the same way, if the unit fraction changes to one-thirteenth, then the number of parts required to make a whole becomes thirteen. An illustration of this is seen in the figure below. As the denominator increases, the size of the parts for each fraction decreases. As the bar goes from one-fourth to one-fifth, the size of the bars decreases, but the size of the denominator increases to five. This pattern continues down the diagram as the bars, or value of the fraction, get smaller, the denominator gets larger.

Comparing Fractions

Comparing fractions requires the use of a common denominator. This necessity can be seen by the two pies below. The first pie has a shaded value of $\frac{2}{10}$ because two pieces are shaded out of the total of ten equal pieces. The second pie has a shaded value of $\frac{2}{7}$ because two pieces are shaded out of a total of seven equal pieces. These two fractions, two-tenths and two-sevenths, have the same numerator and so a misconception may be that they are equal. By looking at the shaded region in each pie, it is apparent that the fractions are not equal. The numerators are the same, but the denominators are not.

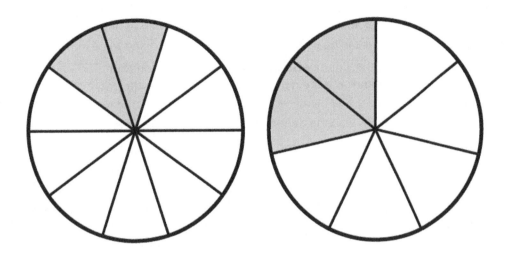

Two parts of a whole are not equivalent unless the whole is broken into the same number of parts. To compare the shaded regions, the denominators seven and ten must be made equal. The lowest number that the two denominators will both divide evenly into is 70, which is the lowest common denominator. Then the numerators must be converted by multiplying by the opposite denominator. These operations result in the two fractions $\frac{14}{70}$ and $\frac{20}{70}$. Now that these two have the same denominator, the conclusion can be made that $\frac{2}{7}$ represents a larger portion of the pie, as seen in the figure below.

Solving Problems Using the Order of Operations

The **order of operations** refers to the order in which problems are to be solved, from parenthesis or grouping, to addition and subtraction. A common way of remembering the order of operations is PEMDAS, or "Please Excuse My Dear Aunt Sally." The letters stand for parenthesis, or grouping, exponents, multiply/divide, and add/subtract. The first step is to complete any operations inside the grouping symbols, or parenthesis. The next step is to simplify all exponents. After exponents, the operations of multiplication and division are performed in the order they appear from left to right. The last operations are addition and subtraction, also performed from left to right. The following problem requires the use of order of operations to be solved:

$$2(3^3 + 5) - 8$$

The first step is to perform the operations inside the grouping symbols, or parenthesis. Inside the parenthesis, the exponent would be performed first, then the addition of $(3^3 + 5)$ which is $(27 + 5)$ or (32). These operations lead to the next step of $2(32) - 8$, where the multiplication can be performed

35

between 2 and 8. This step leads to the problem $64 - 8$, where the answer is 56. The order of operations is important because if solved in a different order, the resulting number would not be 56. A common of when the order of operations can be used is when a store is having a sale and customers may use coupons. Other places may be at a restaurant, for the check, or the gas station when using a card to pay.

Converting Between Fractions, Decimals, and Percents

Within the number system, different forms of numbers can be used. It is important to be able to recognize each type, as well as work with, and convert between, the given forms. The **real number system** comprises natural numbers, whole numbers, integers, rational numbers, and irrational numbers. Natural numbers, whole numbers, integers, and irrational numbers typically are not represented as fractions, decimals, or percentages. Rational numbers, however, can be represented as any of these three forms. A **rational number** is a number that can be written in the form $\frac{a}{b}$, where a and b are integers, and b is not equal to zero. In other words, rational numbers can be written in a fraction form. The value a is the **numerator**, and b is the **denominator.** If the numerator is equal to zero, the entire fraction is equal to zero. Non-negative fractions can be less than 1, equal to 1, or greater than 1. Fractions are less than 1 if the numerator is smaller (less than) than the denominator. For example, $\frac{3}{4}$ is less than 1. A fraction is equal to 1 if the numerator is equal to the denominator. For instance, $\frac{4}{4}$ is equal to 1. Finally, a fraction is greater than 1 if the numerator is greater than the denominator: the fraction $\frac{11}{4}$ is greater than 1.

When the numerator is greater than the denominator, the fraction is called an **improper fraction**. An improper fraction can be converted to a **mixed number,** a combination of both a whole number and a fraction. To convert an improper fraction to a mixed number, divide the numerator by the denominator. Write down the whole number portion, and then write any remainder over the original denominator. For example, $\frac{11}{4}$ is equivalent to $2\frac{3}{4}$. Conversely, a mixed number can be converted to an improper fraction by multiplying the denominator by the whole number and adding that result to the numerator.

Fractions can be converted to decimals. With a calculator, a fraction is converted to a decimal by dividing the numerator by the denominator. For example:

$$\frac{2}{5} = 2 \div 5 = 0.4$$

Sometimes, rounding might be necessary. Consider:

$$\frac{2}{7} = 2 \div 7 = 0.28571429$$

This decimal could be rounded for ease of use, and if it needed to be rounded to the nearest thousandth, the result would be 0.286. If a calculator is not available, a fraction can be converted to a decimal manually. First, find a number that, when multiplied by the denominator, has a value equal to 10, 100, 1,000, etc. Then, multiply both the numerator and denominator times that number. The decimal form of the fraction is equal to the new numerator with a decimal point placed as many place values to the left as there are zeros in the denominator. For example, to convert $\frac{3}{5}$ to a decimal, multiply both the numerator and denominator times 2, which results in $\frac{6}{10}$. The decimal is equal to 0.6 because there is one zero in the denominator, and so the decimal place in the numerator is moved one unit to the left. In the case where rounding would be necessary while working without a calculator, an approximation must be found. A number close to 10, 100, 1,000, etc. can be used. For example, to convert $\frac{1}{3}$ to a decimal, the numerator and denominator can be multiplied by 33 to turn the denominator into approximately 100, which makes for an easier conversion to the equivalent decimal. This process results in $\frac{33}{99}$ and an approximate decimal of 0.33. Once in decimal form, the number can be converted to a percentage. Multiply the decimal by 100 and then place a percent sign after the number. For example, 0.614 is equal to 61.4%. In other words, move the decimal place two units to the right and add the percentage symbol.

Ratios and Rates of Change

Recall that a ratio is the comparison of two different quantities. Comparing 2 apples to 3 oranges results in the ratio 2:3, which can be expressed as the fraction $\frac{2}{5}$. Note that order is important when discussing ratios. The number mentioned first is the antecedent, and the number mentioned second is the consequent. Note that the consequent of the ratio and the denominator of the fraction are *not* the same. When there are 2 apples to 3 oranges, there are five fruit total; two fifths of the fruit are apples, while three fifths are oranges. The ratio 2:3 represents a different relationship that the ratio 3:2. Also, it is important to make sure that when discussing ratios that have units attached to them, the two quantities use the same units. For example, to think of 8 feet to 4 yards, it would make sense to convert 4 yards to feet by multiplying by 3. Therefore, the ratio would be 8 feet to 12 feet, which can be expressed as the fraction $\frac{8}{20}$. Also, note that it is proper to refer to ratios in lowest terms. Therefore, the ratio of 8 feet to 4 yards is equivalent to the fraction $\frac{2}{5}$.

Many real-world problems involve ratios. Often, problems with ratios involve proportions, as when two ratios are set equal to find the missing amount. However, some problems involve deciphering single ratios. For example, consider an amusement park that sold 345 tickets last Saturday. If 145 tickets were sold to adults and the rest of the tickets were sold to children, what would the ratio of the number of adult tickets to children's tickets be? A common mistake would be to say the ratio is 145:345. However, 345 is the total number of tickets sold, not the number of children's tickets. There were $345 - 145 = 200$ tickets sold to children. The correct ratio of adult to children's tickets is 145:200. As a fraction, this expression is written as $\frac{145}{345}$, which can be reduced to $\frac{29}{69}$.

While a ratio compares two measurements using the same units, **rates** compare two measurements with different units. Examples of rates would be $200 for 8 hours of work, or 500 miles traveled per 20 gallons. Because the units are different, it is important to always include the units when discussing rates. Rates can be easily seen because if they are expressed in words, the two quantities are usually split up using one of the following words: *for, per, on, from, in*. Just as with ratios, it is important to write rates in lowest terms. A common rate that can be found in many real-life situations is cost per unit. This quantity describes how much one item or one unit costs. This rate allows the best buy to be determined, given a couple of different sizes of an item with different costs. For example, if 2 quarts of soup was sold for $3.50 and 3 quarts was sold for $4.60, to determine the best buy, the cost per quart should be found. $\frac{\$3.50}{2 \text{ qt}} = \$1.75$ per quart, and $\frac{\$4.60}{3 \text{ qt}} = \$1.53$ per quart. Therefore, the better deal would be the 3-quart option.

Rate of change problems involve calculating a quantity per some unit of measurement. Usually the unit of measurement is time. For example, meters per second is a common rate of change. To calculate this measurement, find the distance traveled in meters and divide by total time traveled. The calculation is an average of the speed over the entire time interval. Another common rate of change used in the real world is miles per hour. Consider the following problem that involves calculating an average rate of change in temperature. Last Saturday, the temperature at 1:00 a.m. was 34 degrees Fahrenheit, and at noon, the temperature had increased to 75 degrees Fahrenheit. What was the average rate of change over that time interval? The average rate of change is calculated by finding the total change in temperature and dividing by the total hours elapsed. Therefore, the rate of change was equal to:

$$\frac{75-34}{12-1} = \frac{41}{11} \text{ degrees per hour}$$

This quantity, rounded to two decimal places, is equal to 3.72 degrees per hour.

A common rate of change that appears in algebra is the slope calculation. Given a linear equation in one variable, $y = mx + b$, the **slope**, m, is equal to $\frac{rise}{run}$ or $\frac{chang\ in\ y}{chang\ in\ x}$. In other words, slope is equivalent to the ratio of the vertical and horizontal changes between any two points on a line. The vertical change is known as the **rise**, and the horizontal change is known as the **run**. Given any two points on a line (x_1, y_1) and (x_2, y_2), slope can be calculated with the formula:

$$m = \frac{y_2 - y_1}{x_2 - x_1} = \frac{\Delta y}{\Delta x}$$

Common real-world applications of slope include determining how steep a staircase should be, calculating how steep a road is, and determining how to build a wheelchair ramp.

Many times, problems involving rates and ratios involve proportions. A proportion states that two ratios (or rates) are equal. The property of cross products can be used to determine if a proportion is true, meaning both ratios are equivalent. If $\frac{a}{b} = \frac{c}{d}$, then to clear the fractions, multiply both sides by the least common denominator, bd. This results in $ad = bc$, which is equal to the result of multiplying along both diagonals. For example, $\frac{4}{40} = \frac{1}{10}$ grants the cross product $4 \times 10 = 40 \times 1$, which is equivalent to $40 = 40$ and shows that this proportion is true. Cross products are used when proportions are involved in real-world problems. Consider the following: If 3 pounds of fertilizer will cover 75 square feet of grass, how many pounds are needed for 375 square feet? To solve this problem, a proportion can be set up using two ratios. Let x equal the unknown quantity, pounds needed for 375 feet. Then, the equation found by setting the two given ratios equal to one another is $\frac{3}{75} = \frac{x}{375}$. Cross-multiplication gives $3 \times 375 = 75x$. Therefore, $1,125 = 75x$. Divide both sides by 75 to get $x = 15$. Therefore, 15 pounds of fertilizer are needed to cover 375 square feet of grass.

Another application of proportions involves similar triangles. If two triangles have the same measurement as two triangles in another triangle, the triangles are said to be **similar.** If two are the same, the third pair of angles are equal as well because the sum of all angles in a triangle is equal to 180 degrees. Each pair of equivalent angles are known as **corresponding angles. Corresponding sides** face the corresponding angles, and it is true that corresponding sides are in proportion.

For example, consider the following set of similar triangles:

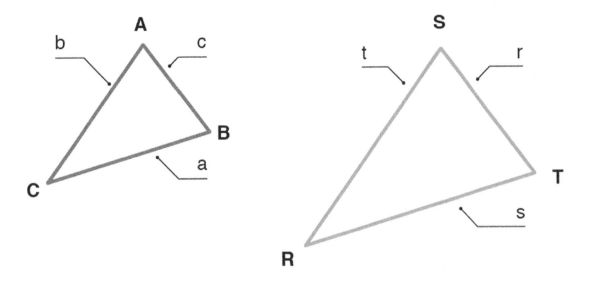

Angles A and S have the same measurement, angles C and R have the same measurement, and angles B and T have the same measurement. Therefore, the following proportion can be set up from the sides:

$$\frac{c}{r} = \frac{a}{s} = \frac{b}{t}$$

This proportion can be helpful in finding missing lengths in pairs of similar triangles. For example, if the following triangles are similar, a proportion can be used to find the missing side lengths, a and b.

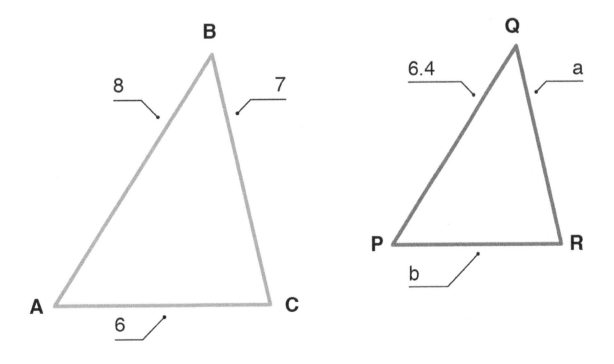

The proportions $\frac{8}{6.4} = \frac{6}{b}$ and $\frac{8}{6.4} = \frac{7}{a}$ can both be cross multiplied and solved to obtain $a = 5.6$ and $b = 4.8$.

A real-life situation that uses similar triangles involves measuring shadows to find heights of unknown objects. Consider the following problem: A building casts a shadow that is 120 feet long, and at the same time, another building that is 80 feet high casts a shadow that is 60 feet long. How tall is the first building? Each building, together with the sun rays and shadows casted on the ground, forms a triangle. They are similar because each building forms a right angle with the ground, and the sun rays form equivalent angles. Therefore, these two pairs of angles are both equal. Because all angles in a triangle add up to 180 degrees, the third angles are equal as well. Both shadows form corresponding sides of the triangle, the buildings form corresponding sides, and the sun rays form corresponding sides. Therefore, the triangles are similar, and the following proportion can be used to find the missing building length:

$$\frac{120}{x} = \frac{60}{80}$$

Cross-multiply to obtain the cross products, $9600 = 60x$. Then, divide both sides by 60 to obtain $x = 160$. This solution means that the other building is 160 feet high.

Percentages

Percentages are defined to be parts per one hundred. To convert a decimal to a percentage, move the decimal point two units to the right and place the percent sign after the number. Percentages appear in many scenarios in the real world. It is important to make sure the statement containing the percentage is translated to a correct mathematical expression. Be aware that it is extremely common to make a mistake when working with percentages within word problems.

An example of a word problem containing a percentage is the following: 35% of people speed when driving to work. In a group of 5,600 commuters, how many would be expected to speed on the way to their place of employment? The answer to this problem is found by finding 35% of 5,600. First, change the percentage to the decimal 0.35. Then compute the product: $0.35 \times 5,600 = 1,960$. Therefore, it would be expected that 1,960 of those commuters would speed on their way to work based on the data given. In this situation, the word "of" signals to use multiplication to find the answer. Another way percentages are used is in the following problem: Teachers work 8 months out of the year. What percent of the year do they work? To answer this problem, find what percent of 12 the number 8 is, because there are 12 months in a year. Therefore, divide 8 by 12, and convert that number to a percentage:

$$\frac{8}{12} = \frac{2}{3} = 0.66\overline{6}$$

The percentage rounded to the nearest tenth place tells us that teachers work 66.7% of the year. Percentages also appear in real-world application problems involving finding missing quantities like in the following question: 60% of what number is 75? To find the missing quantity, an equation can be used. Let x be equal to the missing quantity. Therefore, $0.60x = 75$. Divide each side by 0.60 to obtain 125. Therefore, 60% of 125 is equal to 75.

Sales tax is an important application relating to percentages because tax rates are usually given as percentages. For example, a city might have an 8% sales tax rate. Therefore, when an item is purchased with that tax rate, the real cost to the customer is 1.08 times the price in the store. For example, a $25 pair of jeans costs the customer:

$$\$25 \times 1.08 = \$27$$

Sales tax rates can also be determined if they are unknown when an item is purchased. If a customer visits a store and purchases an item for $21.44, but the price in the store was $19, they can find the tax rate by first subtracting $21.44 − $19 to obtain $2.44, the sales tax amount. The sales tax is a percentage of the in-store price. Therefore, the tax rate is $\frac{2.44}{19} = 0.128$, which has been rounded to the nearest thousandths place. In this scenario, the actual sales tax rate given as a percentage is 12.8%.

Unit Rate Problems

A **unit rate** is a rate with a denominator of one. It is a comparison of two values with different units where one value is equal to one. Examples of unit rates include 60 miles per hour and 200 words per minute. Problems involving unit rates may require some work to find the unit rate. For example, if Mary travels 360 miles in 5 hours, what is her speed, expressed as a unit rate? The rate can be expressed as the following fraction: $\frac{360\ miles}{5\ hours}$. The denominator can be changed to one by dividing by five. The numerator will also need to be divided by five to follow the rules of equality. This division turns the fraction into $\frac{72\ miles}{1\ hour}$, which can now be labeled as a unit rate because one unit has a value of one. Another type question involves the use of unit rates to solve problems. For example, if Trey needs to read 300 pages and his average speed is 75 pages per hour, will he be able to finish the reading in 5 hours? The unit rate is 75 pages per hour, so the total of 300 pages can be divided by 75 to find the time. After the division, the time it takes to read is four hours. The answer to the question is yes, Trey will finish the reading within 5 hours.

Proportional Relationships

Fractions appear in everyday situations, and in many scenarios, they appear in the real-world as ratios and in proportions. A **ratio** is formed when two different quantities are compared. For example, in a group of 50 people, if there are 33 females and 17 males, the ratio of females to males is 33 to 17. This expression can be written in the fraction form as $\frac{33}{50}$, where the denominator is the sum of females and males, or by using the ratio symbol, 33:17. The order of the number matters when forming ratios. In the same setting, the ratio of males to females is 17 to 33, which is equivalent to $\frac{17}{50}$ or 17:33. A **proportion** is an equation involving two ratios. The equation $\frac{a}{b} = \frac{c}{d}$, or $a:b = c:d$ is a proportion, for real numbers a, b, c, and d. Usually, in one ratio, one of the quantities is unknown, and cross-multiplication is used to solve for the unknown. Consider $\frac{1}{4} = \frac{x}{5}$. To solve for x, cross-multiply to obtain $5 = 4x$. Divide each side by 4 to obtain the solution $x = \frac{5}{4}$. It is also true that percentages are ratios in which the second term is 100 minus the first term. For example, 65% is 65:35 or $\frac{65}{100}$. Therefore, when working with percentages, one is also working with ratios.

Real-world problems frequently involve proportions. For example, consider the following problem: If 2 out of 50 pizzas are usually delivered late from a local Italian restaurant, how many would be late out of 235 orders? The following proportion would be solved with x as the unknown quantity of late pizzas:

$$\frac{2}{50} = \frac{x}{235}$$

Cross multiplying results in $470 = 50x$. Divide both sides by 50 to obtain $x = \frac{470}{50}$, which in lowest terms is equal to $\frac{47}{5}$. In decimal form, this improper fraction is equal to 9.4. Because it does not make sense to answer this question with decimals (portions of pizzas do not get delivered) the answer must be rounded.

Traditional rounding rules would say that 9 pizzas would be expected to be delivered late. However, to be safe, rounding up to 10 pizzas out of 235 would probably make more sense.

Algebra

Algebraic Expressions and Equations

An **algebraic expression** is a mathematical phrase that may contain numbers, variables, and mathematical operations. An expression represents a single quantity. For example, $3x + 2$ is an algebraic expression.

An **algebraic equation** is a mathematical sentence with two expressions that are equal to each other. That is, an equation must contain an equals sign, as in $3x + 2 = 17$. This statement says that the value of the expression on the left side of the equals sign is equivalent to the value of the expression on the right side. In an expression, there are not two sides because there is no equals sign. The equals sign ($=$) is the difference between an expression and an equation.

Inequalities look like equations, but instead of an equals sign, $<, >, \leq, \geq$, or $\neq$ are used. Here are some examples of inequalities: $2x + 7 < y, 3x^2 \geq 5$, and $x \neq 4$. Inequalities show relationships between algebraic expressions when the quantities are different. Inequalities can also be expressed in function form if they are solved for y. For instance, the first inequality listed above can be written as:

$$2x + y < f(x)$$

Two algebraic expressions are equivalent if they represent the same expression. Therefore, plugging in the same values into the variables in each expression will result in the same answer. To obtain an equivalent form of an algebraic expression, laws of algebra must be followed. For instance, addition and multiplication are both commutative and associative. Therefore, terms in an algebraic expression can be added in any order and multiplied in any order. For instance, $4x + 2y$ is equivalent to:

$$2y + 4x \text{ and } y \times 2 + x \times 4$$

Also, the distributive law allows a number to be distributed throughout parentheses, as in the following:

$$a(b + c) = ab + ac$$

The two expressions on both sides of the equals sign are equivalent. Also, collecting like terms is important when working with equivalent forms. The simplest version of an expression is always the one easiest to work with, so all like terms (those with the same variables raised to the same powers) must be combined.

Note that an expression is not an equation, and therefore expressions cannot be multiplied times numbers, divided by numbers, or have numbers added to them or subtracted from them and still have equivalent expressions. These processes can only happen in equations when the same step is performed on both sides of the equals sign.

Parts of Expressions

A **variable** is a symbol used to represent a number. Letters like x, y, and z are often used as variables in algebra.

A **constant** is a number that cannot change its value. For example, 18 is a constant.

A **term** is a constant, variable, or the product of constants and variables. In an expression, terms are separated by $+$ and $-$ signs. Examples of terms are $24x$, -32, and $15xyz$.

Like terms are terms that contain the same variables. For example, $6z$ and $-8z$ are like terms, and $9xy$ and $17xy$ are like terms. Constants, like 23 and 51, are like terms as well.

A **factor** is something that is multiplied by something else. A factor may be a constant, a variable, or a sum of constants or variables.

A **coefficient** is the numerical factor in a term that has a variable. In the term $16x$, the coefficient is 16.

Example: Given the expression, $6x - 12y + 18$, answer the following questions.

How many terms are in the expression?

- Solution: 3

Name the terms.

- Solution: 6x, –12y, and 18   (Notice that the minus sign preceding the 12 is interpreted to represent negative 12)

Name the factors.

- Solution: 6, x, –12, y

What are the coefficients in this expression?

- Solution: 6 and –12

What is the constant in this expression?

- Solution: 18

Adding and Subtracting Linear Algebraic Expressions

To add and subtract linear algebra expressions, you must combine like terms. **Like terms** are described as those terms that have the same variable with the same exponent. In the following example, the x-terms can be added because the variable is the same and the exponent on the variable of one is also the same. These terms add to be $9x$. The other like terms are called **constants** because they have no variable component. These terms will add to be nine.

Example: Add $\quad\quad$ $(3x - 5) + (6x + 14)$

$\quad\quad\quad\quad\quad$ $3x - 5 + 6x + 14$ $\quad\quad\quad\quad\quad\quad$ Rewrite without parentheses

$\quad\quad\quad\quad\quad$ $3x + 6x - 5 + 14$ $\quad\quad\quad\quad\quad\quad$ Commutative property of addition

$\quad\quad\quad\quad\quad$ $9x + 9$ $\quad\quad\quad\quad\quad\quad\quad\quad\quad$ Combine like terms

When subtracting linear expressions, be careful to add the opposite when combining like terms. Do this by distributing -1, which is multiplying each term inside the second parenthesis by negative one. Remember that distributing -1 changes the sign of each term.

Example: Subtract $\quad\quad$ $(17x + 3) - (27x - 8)$

$\quad\quad\quad\quad\quad$ $17x + 3 - 27x + 8$ $\quad\quad\quad\quad\quad\quad$ Distributive Property

$\quad\quad\quad\quad\quad$ $17x - 27x + 3 + 8$ $\quad\quad\quad\quad\quad\quad$ Commutative property of addition

$\quad\quad\quad\quad\quad$ $-10x + 11$ $\quad\quad\quad\quad\quad\quad\quad\quad$ Combine like terms

Example: Simplify by adding or subtracting:

$\quad\quad\quad\quad\quad$ $(6m + 28z - 9) + (14m + 13) - (-4z + 8m + 12)$

$\quad\quad\quad\quad\quad$ $6m + 28z - 9 + 14m + 13 + 4z - 8m - 12$ $\quad\quad$ Distributive Property

$\quad\quad\quad\quad\quad$ $6m + 14m - 8m + 28z + 4z - 9 + 13 - 12$ $\quad\quad$ Commutative Property of Addition

$\quad\quad\quad\quad\quad$ $12m + 32z - 8$ $\quad\quad\quad\quad\quad\quad\quad\quad\quad$ Combine like terms

The Distributive Property

The Distributive Property: $a(b + c) = ab + ac$

The **distributive property** is a way of taking a factor and multiplying it through a given expression in parentheses. Each term inside the parentheses is multiplied by the outside factor, eliminating the parentheses. The following example shows how to distribute the number 3 to all the terms inside the parentheses.

Example: Use the distributive property to write an equivalent algebraic expression:

$\quad\quad\quad\quad\quad$ $3(2x + 7y + 6)$

$\quad\quad\quad\quad\quad$ $3(2x) + 3(7y) + 3(6)$ $\quad\quad\quad\quad\quad\quad$ Distributive property

$\quad\quad\quad\quad\quad$ $6x + 21y + 18$ $\quad\quad\quad\quad\quad\quad\quad\quad$ Simplify

Because $a - b$ can be written $a + (-b)$, the distributive property can be applied in the example below.

Example: Use the distributive property to write an equivalent algebraic expression.

        $7(5m - 8)$

        $7[5m + (-8)]$ Rewrite subtraction as addition of -8

        $7(5m) + 7(-8)$ Distributive property

        $35m - 56$ Simplify

In the following example, note that the factor of 2 is written to the right of the parentheses but is still distributed as before.

Example: Use the distributive property to write an equivalent algebraic expression:

        $(3m + 4x - 10)2$

        $(3m)2 + (4x)2 + (-10)2$ Distributive property

        $6m + 8x - 20$ Simplify

Example: $-(-2m + 6x)$

In this example, the negative sign in front of the parentheses can be interpreted as $-1(-2m + 6x)$

        $-1(-2m + 6x)$

        $-1(-2m) + (-1)(6x)$ Distributive property

        $2m - 6x$ Simplify

Evaluating Simple Algebraic Expressions for Given Values of Variables

To evaluate an algebra expression for a given value of a variable, replace the variable with the given value. Then perform the given operations to simplify the expression.

Example: Evaluate $12 + x$ for $x = 9$

        $12 + (9)$ Replace x with the value of 9 as given in the problem. It is a good idea to always use parentheses when substituting this value. This will be particularly important in the following examples.

        21   Add

Now see that when x is 9, the value of the given expression is 21.

Example: Evaluate $4x + 7$ for $x = 3$

$$4(3) + 7 \qquad \text{Replace the } x \text{ in the expression with 3}$$

$$12 + 7 \qquad \text{Multiply (remember order of operations)}$$

$$19 \qquad \text{Add}$$

Therefore, when x is 3, the value of the given expression is 19.

Example: Evaluate $-7m - 3r - 18$ for $m = 2$ and $r = -1$

$$-7(2) - 3(-1) - 18 \qquad \text{Replace } m \text{ with 2 and } r \text{ with -1}$$

$$-14 + 3 - 18 \qquad \text{Multiply}$$

$$-29 \qquad \text{Add}$$

So, when m is 2 and r is -1, the value of the given expression is -29.

Solving Real-World Single- or Multistep Equations or Inequalities that Contain an Unknown

When presented with a real-world problem that must be solved, the first step is always to determine what the unknown quantity is that must be solved for. Use a variable, such as x or t, to represent that unknown quantity. Sometimes, there can be two or more unknown quantities. In this case, either choose an additional variable, or if a relationship exists between the unknown quantities, express the other quantities in terms of the original variable. After choosing the variables, form algebraic expressions and/or equations that represent the verbal statement in the problem. The following table shows examples of vocabulary used to represent the different operations:

Addition	Sum, plus, total, increase, more than, combined, in all
Subtraction	Difference, less than, subtract, reduce, decrease, fewer, remain
Multiplication	Product, multiply, times, part of, twice, triple
Division	Quotient, divide, split, each, equal parts, per, average, shared

The combination of operations and variables form both mathematical expression and equations. As mentioned, the difference between expressions and equations are that there is no equals sign in an expression, and that expressions are **evaluated** to find an unknown quantity, while equations are **solved** to find an unknown quantity. Also, inequalities can exist within verbal mathematical statements. Instead of a statement of equality, expressions state quantities are *less than, less than or equal to, greater than,* or *greater than or equal to.* Another type of inequality is when a quantity is said to be *not equal to* another quantity. The symbol used to represent "not equal to" is $\neq$.

The steps for solving inequalities in one variable are the same steps for solving equations in one variable. The addition and multiplication principles are used. However, to maintain a true statement when using the $<$, $\leq$, $>$, and $\geq$ symbols, if a negative number is either multiplied by both sides of an inequality or divided from both sides of an inequality, the sign must be flipped. For instance, consider the following inequality: $3 - 5x \leq 8$. First, 3 is subtracted from each side to obtain $-5x \leq 5$. Then, both sides are divided by -5, while flipping the sign, to obtain $x \geq -1$. Therefore, any real number greater than or equal to -1 satisfies the original inequality.

Real-world problems can be translated into both one-step and multi-step problems. In either case, the word problem must be translated from the verbal form into mathematical expressions and equations that can be solved using algebra. An example of a one-step real-world problem is the following: A cat weighs half as much as a dog living in the same house. If the dog weighs 14.5 pounds, how much does the cat weigh? To solve this problem, an equation can be used. In any word problem, the first step is to define variables that represent the unknown quantities. For this problem, let x be equal to the unknown weight of the cat. Because two times the weight of the cat equals 14.5 pounds, the equation to be solved is: $2x = 14.5$. Use the multiplication principle to divide both sides by 2. Therefore, $x = 7.25$. The cat weighs 7.25 pounds.

Most of the time, real-world problems are more difficult than this one and consist of multi-step problems. The following is an example of a multi-step problem: The sum of two consecutive page numbers is equal to 437. What are those page numbers? First, define the unknown quantities. If x is equal to the first page number, then $x + 1$ is equal to the next page number because they are consecutive integers. Their sum is equal to 437, and this statement translates to the equation:

$$x + x + 1 = 437$$

To solve, first collect like terms to obtain:

$$2x + 1 = 437$$

Then, subtract 1 from both sides and then divide by 2. The solution to the equation is $x = 218$. Therefore, the two consecutive page numbers that satisfy the problem are 218 and 219. It is always important to make sure that answers to real-world problems make sense. For instance, it should be a red flag if the solution to this same problem resulted in decimals, which would indicate the need to check the work. Page numbers are whole numbers; therefore, if decimals are found to be answers, the solution process should be double-checked to see where mistakes were made.

Evaluating Simple Formulas and Expressions

Given the formula for the area of a rectangle $A = lw$, with A = area, l = length, and w = width, the area of a rectangle can be determined, given the length and the width.

For example, if the length of a rectangle is 7 cm and the width is 10 cm, find the area of the rectangle.

Solution: Just as when evaluating expressions, replace the variables with the given values.

Given $A = lw$, $l = 7$ and $w = 10$.

$A = (7)(10)$ Replace l with 7 and w with 10

$A = 70$ Multiply

Therefore, the area of the rectangle is 70 cm^2.

Example: The formula for perimeter of a rectangle, $P = 2l + 2w$, where P is perimeter, l is length, and w is width. If the length of a rectangle is 12 inches and the width is 9 inches, find the perimeter.

Solution:

$$P = 2l + 2w$$

$$P = 2(12) + 2(9)$$ Replace l with 12 and w with 9

$$P = 24 + 18$$ Use correct order of operations; multiply first

$$P = 42$$ Add

The perimeter of this rectangle is 42 inches.

Independent variables are independent, meaning they are not changed by other variables within the context of the problem. **Dependent variables** are dependent, meaning they may change depending on how other variables change in the problem. For example, in the formula for the perimeter of a fence, the length and width are the independent variables and the perimeter is the dependent variable. The formula is shown below.

$$P = 2l + 2w$$

As the width or the length changes, the perimeter may also change. The first variables to change are the length and width, which then result in a change in perimeter. The change does not come first with the perimeter and then with length and width. When comparing these two types of variables, it is helpful to ask which variable causes the change and which variable is affected by the change.

Another formula to represent this relationship is the formula for circumference shown below.

$$C = \pi \times d$$

The C represents circumference and the d represents diameter. The pi symbol is approximated by the fraction $\frac{22}{7}$, or 3.14. In this formula, the diameter of the circle is the independent variable. It is the portion of the circle that changes, which changes the circumference as a result. The circumference is the variable that is being changed by the diameter, so it is called the dependent variable. It depends on the value of the diameter.

Another place to recognize independent and dependent variables can be in experiments. A common experiment is one where the growth of a plant is tested based on the amount of sunlight it receives. Each plant in the experiment is given a different amount of sunlight, but the same amount of other nutrients like light and water. The growth of the plants is measured over a given time period and the results show how much sunlight is best for plants. In this experiment, the independent variable is the amount of sunlight that each plant receives. The dependent variable is the growth of each plant. The growth depends on the amount of sunlight, which gives reason for the distinction between independent and dependent variables.

Linear Relationships

Linear growth involves a quantity, the **dependent variable**, increasing or decreasing at a constant rate as another quantity, the **independent variable**, increases as well. The graph of linear growth is a straight line. Linear growth is represented as the following equation: $y = mx + b$, where m is the **slope** of the line, also known as the **rate of change**, and b is the **y-intercept**. If the y-intercept is 0, then the linear growth is actually known as **direct variation**. If the slope is positive, the dependent variable increases as the independent variable increases, and if the slope is negative, the dependent variable decreases as the independent variable increases.

A linear function that models a linear relationship between two quantities is of the form $y = mx + b$, or in function form $f(x) = mx + b$. In a linear function, the value of y depends on the value of x, and y increases or decreases at a constant rate as x increases. Therefore, the independent variable is x, and the dependent variable is y. The graph of a linear function is a line, and the constant rate can be seen by looking at the steepness, or **slope**, of the line. If the line increases from left to right, the slope is positive. If the line slopes downward from left to right, the slope is negative.

Slope is rise over run or how much the y-values change for a given change in x-values. *In* the function, m represents slope. Each point on the line is an **ordered pair** (x, y), where x represents the x-coordinate of the point and y represents the y-coordinate of the point. The point where $x = 0$ is known as the y-intercept, and it is the place where the line crosses the y-axis. If $x = 0$ is plugged into $f(x) = mx + b$, the result is $f(0) = b$, so therefore, the point $(0, b)$ is the y-intercept of the line. The derivative of a linear function is its slope. The slope can also be determined by finding the difference in the y-values between two ordered pairs and dividing this difference by the difference in the x-values of the same two ordered pairs.

Consider the following situation. A taxicab driver charges a flat fee of $2 per ride and $3 a mile. This statement can be modeled by the function $f(x) = 3x + 2$ where x represents the number of miles and $f(x) = y$ represents the total cost of the ride. The total cost increases at a constant rate of $2 per mile, and that is why this situation is a linear relationship. The slope $m = 3$ is equivalent to this rate of change. The flat fee of $2 is the y-intercept. It is the place where the graph crosses the x-axis, and it represents the cost when $x = 0$, or when no miles have been traveled in the cab. The y-intercept in this situation represents the flat fee.

Two lines are **parallel** if they never intersect. Given the equation of two lines, they are parallel if they have the same slope and different y-intercepts. If they had the same slope and same y-intercept, they would be the same line. Therefore, in order to show two lines are parallel, put them in slope-intercept form, $y = mx + b$, to find m and b. The two lines $y = 2x + 6$ and $4x - 2y = 6$ are parallel. The second line in slope intercept is:

$$y = 2x - 3$$

Both lines have the same slope, 2, and different y-intercepts.

Two lines are **perpendicular** if they intersect at a right angle. Given the equation of two lines, they are perpendicular if their slopes are negative reciprocals. Therefore, the product of both slopes is equal to $- 1$. For example, the lines $y = 4x + 1$ and $y = -\frac{1}{4}x + 1$ are perpendicular because their slopes are negative reciprocals. The product of 4 and $-\frac{1}{4}$ is -1.

Solving Equations in One Variable

An **equation in one variable** is a mathematical statement where two algebraic expressions in one variable, usually x, are set equal. To solve the equation, the variable must be isolated on one side of the equals sign. The addition and multiplication principles of equality are used to isolate the variable. The **addition principle of equality** states that the same number can be added to or subtracted from both sides of an equation. Because the same value is being used on both sides of the equals sign, equality is maintained. For example, the equation $2x = 5x$ is equivalent to both $2x + 3 = 5x + 3$, and $2x - 5 = 5x - 5$. This principle can be used to solve the following equation: $x + 5 = 4$. The variable x must be isolated, so to move the 5 from the left side, subtract 5 from both sides of the equals sign. Therefore:

$$x + 5 - 5 = 4 - 5$$

So, the solution is $x = -1$. This process illustrates the idea of an **additive inverse** because subtracting 5 is the same as adding -5. Basically, add the opposite of the number that must be removed to both sides of the equals sign. The **multiplication principle of equality** states that equality is maintained when a number is either multiplied times both expressions on each side of the equals sign, or when both expressions are divided by the same number. For example, $4x = 5$ is equivalent to both $16x = 20$ and $x = \frac{5}{4}$. Multiplying both sides times 4 and dividing both sides by 4 maintains equality. Solving the equation $6x - 18 = 5$ requires the use of both principles. First, apply the addition principle to add 18 to both sides of the equals sign, which results in $6x = 23$. Then use the multiplication principle to divide both sides by 6, giving the solution $x = \frac{23}{6}$. Using the multiplication principle in the solving process is the same as involving a multiplicative inverse. A **multiplicative inverse** is a value that, when multiplied by a given number, results in 1. Dividing by 6 is the same as multiplying by $\frac{1}{6}$, which is both the reciprocal and multiplicative inverse of 6.

When solving a linear equation in one variable, checking the answer shows if the solution process was performed correctly. Plug the solution into the variable in the original equation. If the result is a false statement, something was done incorrectly during the solution procedure. Checking the example above gives the following:

$$6 \times \frac{23}{6} - 18 = 23 - 18 = 5$$

Therefore, the solution is correct.

Some equations in one variable involve fractions or the use of the distributive property. In either case, the goal is to obtain only one variable term and then use the addition and multiplication principles to isolate that variable. Consider the equation $\frac{2}{3}x = 6$. To solve for x, multiply each side of the equation by the reciprocal of $\frac{2}{3}$, which is $\frac{3}{2}$. This step results in $\frac{3}{2} \times \frac{2}{3}x = \frac{3}{2} \times 6$, which simplifies into the solution $x = 9$. Now consider the equation:

$$3(x + 2) - 5x = 4x + 1$$

Use the distributive property to clear the parentheses. Therefore, multiply each term inside the parentheses by 3. This step results in:

$$3x + 6 - 5x = 4x + 1$$

Next, collect like terms on the left-hand side. **Like terms** are terms with the same variable or variables raised to the same exponent(s). Only like terms can be combined through addition or subtraction. After collecting like terms, the equation is:

$$-2x + 6 = 4x + 1$$

Finally, apply the addition and multiplication principles. Add $2x$ to both sides to obtain:

$$6 = 6x + 1$$

Then, subtract 1 from both sides to obtain $5 = 6x$. Finally, divide both sides by 6 to obtain the solution:

$$\frac{5}{6} = x$$

Two other types of solutions can be obtained when solving an equation in one variable. The final result could be that there is either no solution or that the solution set contains all real numbers. Consider the equation:

$$4x = 6x + 5 - 2x$$

First, the like terms can be combined on the right to obtain:

$$4x = 4x + 5$$

Next, subtract $4x$ from both sides. This step results in the false statement $0 = 5$. There is no value that can be plugged into x that will ever make this equation true. Therefore, there is no solution. The solution procedure contained correct steps, but the result of a false statement means that no value satisfies the equation. The symbolic way to denote that no solution exists is $\emptyset$. Next, consider the equation:

$$5x + 4 + 2x = 9 + 7x - 5$$

Combining the like terms on both sides results in:

$$7x + 4 = 7x + 4$$

The left-hand side is exactly the same as the right-hand side. Using the addition principle to move terms, the result is $0 = 0$, which is always true. Therefore, the original equation is true for any number, and the solution set is all real numbers. The symbolic way to denote such a solution set is $\mathbb{R}$, or in interval notation, $(-\infty, \infty)$.

Solving Equations Using Reasoning

The reasonableness of an answer found in a math problem gives evidence to the accuracy of the work. If the answer is not reasonable, the work should be redone in order to find the error and correct the problem. Problems that involve fractions and decimals are good places to use reasonableness to check answers. For example, Karen has $63.75 to spend on sodas for her family gathering. If each soda costs $1.50, how many can she buy? The answer can be found by division, but because there are decimals, an estimate can be found by rounding the two numbers and doing easy division. The money can round to $64 and the sodas can round to $2. An estimate is 32 sodas. When the actual division is done, the answer should be close to 32. If not, it is a sign that there is an error in the math.

Series and Sequences

A **sequence of numbers** is a list of numbers that follows a specific pattern. Each member of the sequence is known as an **individual term** of the sequence, and a formula can be found to represent each term. For example, the list of numbers 5, 10, 15, 20, ... is a sequence of numbers, and ... shows that the sequence continues indefinitely. Each term represents a multiple of 5. The first term is 1×5, the second term is

2×5, the third term is 3×5, etc. In general, the nth term is $5 \times n$. Other, more complicated sequences can exist as well. For example, each term of the sequence 1, 4, 9, 16, 25, ... is not found through addition or multiplication. Each term happens to be a perfect square, and the first term is 1 squared, the second term is 2 squared, etc. Therefore, the nth term is n^2. A famous sequence of numbers is the Fibonacci sequence, which consists of 0, 1, 1, 2, 3, 5, 8, 13, 21, ... After the first two terms 0 and 1, all terms are found by adding the two previous terms. Therefore, the next term in the sequence is $13 + 21 = 34$. The formula for the nth term is defined recursively, using the two previous terms.

Number and Shape Patterns

Patterns in math are those sets of numbers or shapes that follow a rule. Given a set of values, patterns allow the question of "what's next?" to be answered. In the following set there are two types of shapes, a lighter rectangle and a darker circle. The set contains a pattern because every odd-placed shape is a lighter rectangle and every even-placed spot is taken by a darker circle. This is a pattern because there is a rule of lighter rectangle, then darker circle, that is followed to find the set.

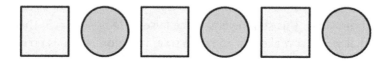

A set of numbers can also be described as having a pattern if there is a rule that can be followed to reproduce the set. The following set of numbers has a rule of adding 3 each time. It begins with zero and increases by 3 each time. By following this rule and pattern, the number after 12 is found to be 15. Further extending the pattern, the numbers are 18, 21, 24, 27. The pattern of increasing by multiples of three can describe this pattern.

A pattern can also be generated from a given rule. Starting with zero, the rule of adding 5 can be used to produce a set of numbers. The following list will result from using the rule: 0, 5, 10, 15, 20. Describing this pattern can include words such as "multiples" of 5 and an "increase" of 5. Any time this pattern needs to be extended, the rule can be applied to find more numbers. Patterns are identified by the rules they follow. This rule should be able to generate new numbers or shapes, while also applying to the given numbers or shapes.

Making Predictions Based on Patterns
Given a certain pattern, future numbers or shapes can be found. Pascal's triangle is an example of a pattern of numbers. Questions can be asked of the triangle, such as, "what comes next?" and "what values determine the next line?" By examining the different parts of the triangle, conjectures can be made about how the numbers are generated. For the first few rows of numbers, the increase is small. Then the numbers begin to increase more quickly. By looking at each row, a conjecture can be made that the sum of the first row determines the second row's numbers. The second row's numbers can be added to find the third row. To test this conjecture, two numbers can be added, and the number found directly between and below them should be that sum. For the third row, the middle number is 2, which is the sum of the two 1s above it. For the fifth row, the 1 and 3 can be added to find a sum of 4, the same number below the 1 and 3. This pattern continues throughout the triangle. Once the pattern is confirmed to apply throughout the triangle, predictions can be made for future rows. The sums of the bottom row numbers

can be found and then added to the bottom of the triangle. In more general terms, the diagonal rows have patterns as well. The outside numbers are always 1. The second diagonal rows are in counting order. The third diagonal row increases each time by one more than the previous. It is helpful to generalize patterns because it makes the pattern more useful in terms of applying it. Pascal's triangle can be used to predict the tossing of a coin, predicting the chances of heads or tails for different numbers of tosses. It can also be used to show the Fibonacci Sequence, which is found by adding the diagonal numbers together.

Pascal's Triangle

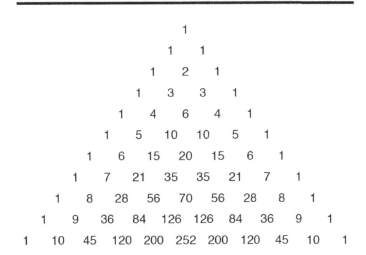

Identifying Relationships Between the Corresponding Terms of Two Numerical Patterns

Sets of numerical patterns can be found by starting with a number and following a given rule. If two sets are generated, the corresponding terms in each set can be found to relate to one another by one or more operations. For example, the following table shows two sets of numbers that each follow their own pattern. The first column shows a pattern of numbers increasing by 1. The second column shows the numbers increasing by 4. Because the numbers are lined up, corresponding numbers are side by side for the two sets. A question to ask is, "How can the number in the first column be turned into the number in the second column?"

1	4
2	8
3	12
4	16
5	20

This answer will lead to the relationship between the two sets. By recognizing the multiples of 4 in the right column and the counting numbers in order in the left column, the relationship of multiplying by four is determined. The first set is multiplied by 4 to get the second set of numbers. To confirm this relationship, each set of corresponding numbers can be checked. For any two sets of numerical patterns, the corresponding numbers can be lined up to find how each one relates to the other. In some cases, the relationship is simply addition or subtraction, multiplication or division. In other relationships, these operations are used in conjunction with each together. As seen in the following table, the relationship uses multiplication and addition. The following expression shows this relationship: $3x + 2$. The x represents the numbers in the first column.

1	5
2	8
3	11
4	14

Geometry

Characteristics of Common Two- and Three-Dimensional Figures

Shapes are defined by their angles and number of sides. A shape with one continuous side, where all points on that side are equidistant from a center point is called a **circle.** A shape made with three straight line segments is a **triangle.** A shape with four sides is called a **quadrilateral,** but more specifically a *square*, **rectangle, parallelogram,** or **trapezoid,** depending on the interior angles. These shapes are two-dimensional and only made of straight lines and angles.

Solids can be formed by combining these shapes and forming three-dimensional figures. These figures have another dimension because they add one more direction. Examples of solids may be prisms or spheres. There are four figures below that can be described based on their sides and dimensions. Figure 1 is a cone because it has three dimensions, where the bottom is a circle and the top is formed by the sides combining to one point. Figure 2 is a triangle because it has two dimensions, made up of three-line segments. Figure 3 is a cylinder made up of two base circles and a rectangle to connect them in three dimensions. Figure 4 is an oval because it is one continuous line in two dimensions, not equidistant from the center.

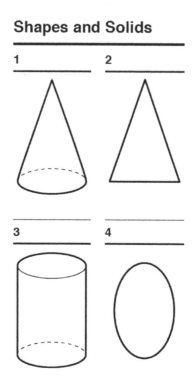

Shapes and Solids

1

2

3

4

Figure 5 below is made up of squares in three dimensions, combined to make a cube. Figure 6 is a rectangle because it has four sides that intersect at right angles. More specifically, it can be described as a **square** because the four sides have equal measures. Figure 7 is a pyramid because the bottom shape is a square and the sides are all triangles. These triangles intersect at a point above the square. Figure 8 is a circle because it is made up of one continuous line where the points are all equidistant from one center point.

Shapes and Solids

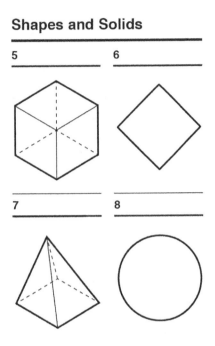

5

6

7

8

Properties of Lines

Geometric figures can be identified by matching the definition with the object. For example, a **line segment** is made up of two endpoints and the line drawn between them. A **ray** is made up of one endpoint and one extending side that goes on forever. A **line** has no endpoints and two sides that extend on forever. These three geometric figures are shown below. What happens at A and B determines the name of each figure.

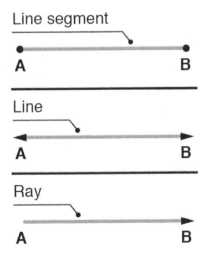

Line segment

A   B

Line

A   B

Ray

A   B

Parallel and perpendicular lines are made up of two lines, like the second figure above. They are distinguished from each other by how the two lines interact. **Parallel** lines run alongside one another, but they never intersect. **Perpendicular** lines intersect at a 90-degree, or a right, angle. An example of these two sets of lines is shown below. Also shown in the figure are non-examples of these two types of lines. Because the first set of lines, in the top left corner, will eventually intersect if they continue, they are not parallel. In the second set, the lines run in the same direction and will never intersect, making them parallel. The third set, in the bottom left corner, intersect at an angle that is not right, or not 90 degrees. The fourth set is perpendicular because the lines intersect at exactly a right angle.

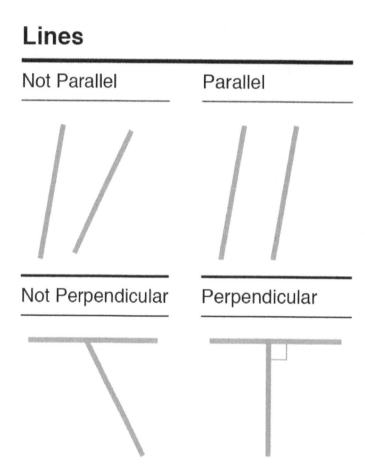

Classifying Angles

When two rays are joined together at their endpoints, an **angle** is formed. Angles can be described based on their measure. An angle whose measure is 90 degrees is described as a right angle, just as with perpendicular lines. Ninety degrees is a standard to which other angles are compared. If an angle is less than ninety degrees, it is an **acute angle**. If it is greater than ninety degrees, it is an **obtuse angle**. If an angle is equal to twice a right angle, or 180 degrees, it is a **straight angle**.

Examples of these types of angles are shown below:

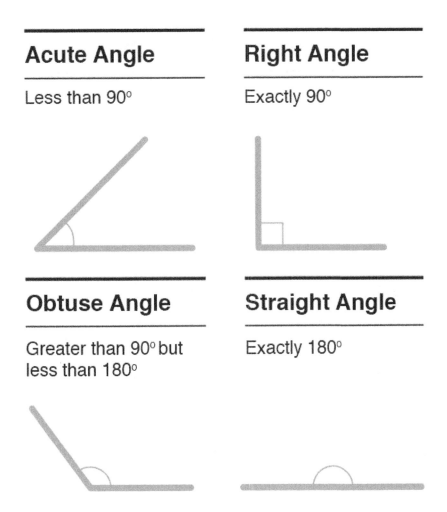

Acute Angle
Less than 90°

Right Angle
Exactly 90°

Obtuse Angle
Greater than 90° but less than 180°

Straight Angle
Exactly 180°

A **straight angle** is equal to 180 degrees, or a straight line. If the line continues through the **vertex,** or point where the rays meet, and does not change direction, then the angle is straight. This is shown in Figure 1 below. The second figure shows an obtuse angle. Its measure is greater than ninety degrees, but less than that of a straight angle. An estimate for its measure may be 175 degrees. Figure 3 shows an acute angle because it is just less than that of a right angle. Its measure may be estimated to be 80 degrees.

The last image, Figure 4, shows another acute angle. This measure is much smaller, at approximately 35 degrees, but it is still classified as acute because it is between zero and 90 degrees.

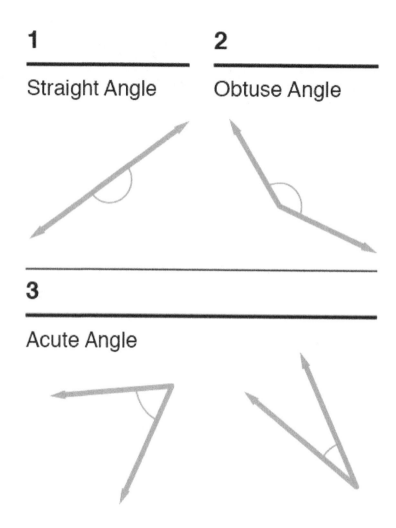

1

Straight Angle

2

Obtuse Angle

3

Acute Angle

Perimeter and Area

Perimeter and area are two commonly used geometric quantities that describe objects. **Perimeter** is the distance around an object. The perimeter of an object can be found by adding the lengths of all sides. Perimeter may be used in problems dealing with lengths around objects such as fences or borders. It may also be used in finding missing lengths or working backwards. If the perimeter is given, but a length is missing, subtraction can be used to find the missing length. Given a square with side length s, the formula for perimeter is $P = 4s$. Given a rectangle with length l and width w, the formula for perimeter is $P = 2l + 2w$. The perimeter of a triangle is found by adding the three side lengths, and the perimeter of a trapezoid is found by adding the four side lengths. The units for perimeter are always the original units of length, such as meters, inches, miles, etc. When discussing a circle, the distance around the object is referred to as its **circumference,** not perimeter. The formula for circumference of a circle is $C = 2\pi r$, where r represents the radius of the circle. This formula can also be written as $C = d\pi$, where d represents the diameter of the circle.

Area is the two-dimensional space covered by an object. These problems may include the area of a rectangle, a yard, or a wall to be painted. Finding the area may be a simple formula, or it may require multiple formulas to be used together. The units for area are square units, such as square meters, square inches, and square miles. Given a square with side length s, the formula for its area is $A = s^2$. Some other common shapes are shown below:

Shape	Formula	Graphic
Rectangle	$Area = length \times width$	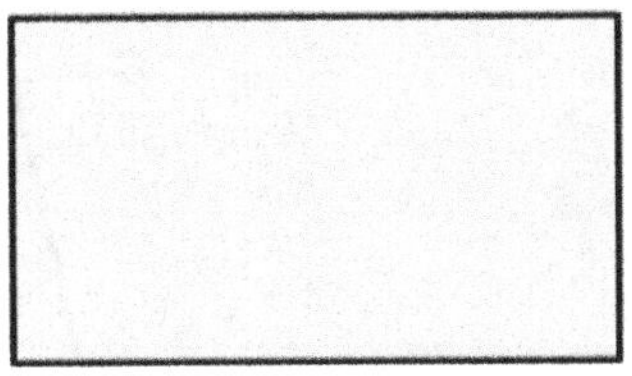
Triangle	$Area = \dfrac{1}{2} \times base \times height$	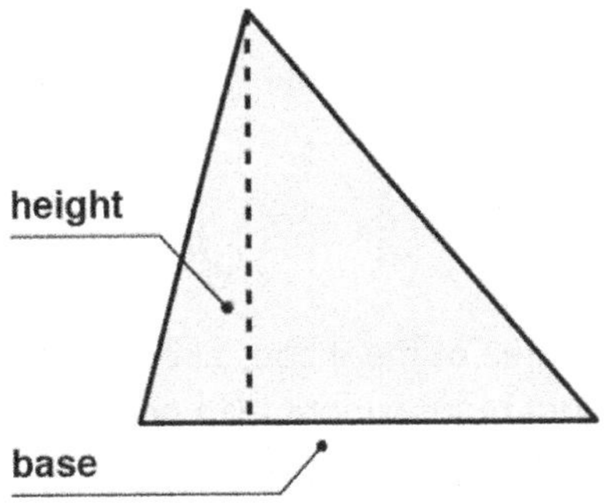
Circle	$Area = \pi \times radius^2$	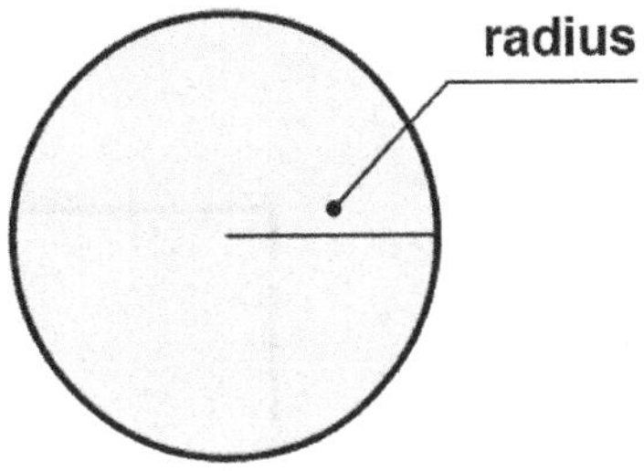

The following formula, not as widely used as those shown above, but very important, is the area of a trapezoid:

Area of a Trapezoid

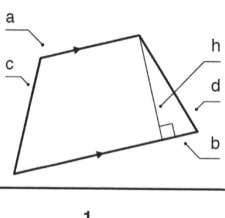

$$A = \frac{1}{2}(a+b)h$$

To find the area of the shapes above, use the given dimensions of the shape in the formula. Complex shapes might require more than one formula. To find the area of the figure below, break the figure into two shapes. The rectangle has dimensions 6 cm by 7 cm. The triangle has dimensions 6 cm by 6 cm. Plug the dimensions into the rectangle formula: $A = 6 \times 7$. Multiplication yields an area of 42 cm². The triangle area can be found using the formula $A = \frac{1}{2} \times 4 \times 6$. Multiplication yields an area of 12 cm². Add the areas of the two shapes to find the total area of the figure, which is 54 cm².

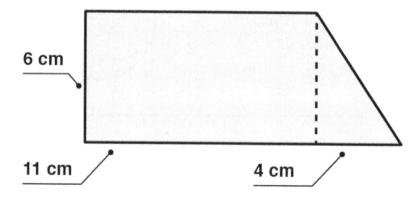

Instead of combining areas, some problems may require subtracting them, or finding the difference.

To find the area of the shaded region in the figure below, determine the area of the whole figure. Then subtract the area of the circle from the whole.

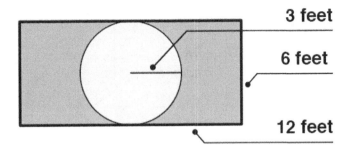

The following formula shows the area of the outside rectangle: $A = 12 \times 6 = 72$ ft^2. The area of the inside circle can be found by the following formula: $A = \pi(3)^2 = 9\pi = 28.3$ ft^2. As the shaded area is outside the circle, the area for the circle can be subtracted from the area of the rectangle to yield an area of 43.7 ft^2.

While some geometric figures may be given as pictures, others may be described in words. If a rectangular playing field with dimensions 95 meters long by 50 meters wide is measured for perimeter, the distance around the field must be found. The perimeter includes two lengths and two widths to measure the entire outside of the field. This quantity can be calculated using the following equation: $P = 2(95) + 2(50) = 290$ m. The distance around the field is 290 meters.

Volume

Perimeter and area are two-dimensional descriptions; volume is three-dimensional. **Volume** describes the amount of space that an object occupies, but it's different from area because it has three dimensions instead of two. The units for volume are cubic units, such as cubic meters, cubic inches, and cubic miles. Volume can be found by using formulas for common objects such as cylinders and boxes.

The following chart shows a diagram and formula for the volume of two objects:

Shape	Formula	Diagram
Rectangular Prism (box)	$V = length \times width \times height$	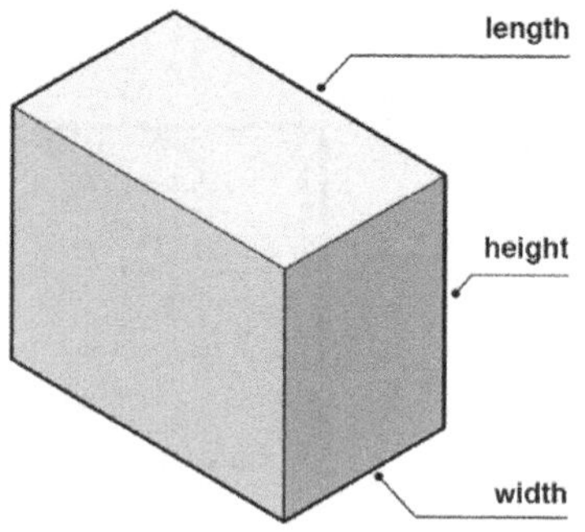
Cylinder	$V = \pi \times radius^2 \times height$	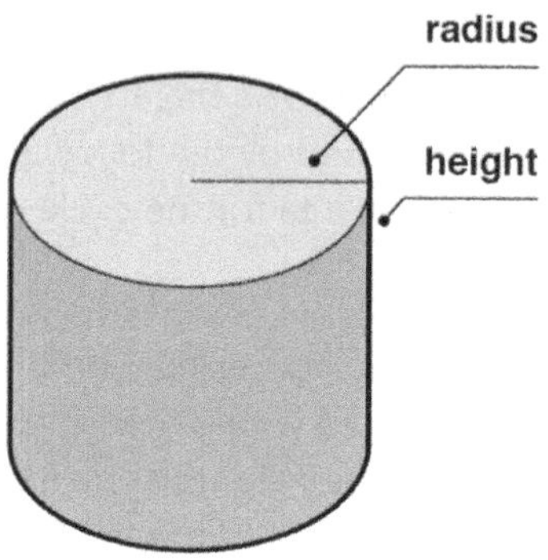

Volume formulas of these two objects are derived by finding the area of the bottom two-dimensional shape, such as the circle or rectangle, and then multiplying times the height of the three-dimensional shape. Other volume formulas include the volume of a cube with side length s: $V = s^3$; the volume of a sphere with radius r: $V = \frac{4}{3}\pi r^3$; and the volume of a cone with radius r and height h:

$$V = \frac{1}{3}\pi r^2 h$$

If a soda can has a height of 5 inches and a radius on the top of 1.5 inches, the volume can be found using one of the given formulas. A soda can is a cylinder. Knowing the given dimensions, the formula can be completed as follows:

$$V = \pi(radius)^2 \times height$$

$$\pi(1.5 \text{ in})^2 \times 5 \text{ in} = 35.325 \text{ in}^3$$

Notice that the units for volume are inches cubed because it refers to the number of cubic inches required to fill the can.

With any geometric calculations, it's important to determine what dimensions are given and what quantities the problem is asking for. If a connection can be made between them, the answer can be found.

Other geometric quantities can include angles inside a triangle. The sum of the measures of three angles in any triangle is 180 degrees. Therefore, if only two angles are known inside a triangle, the third can be found by subtracting the sum of the two known quantities from 180. Two angles whose sum is equal to 90 degrees are known as **complementary angles.** For example, angles measuring 72 and 18 degrees are complementary, and each angle is a complement of the other. Finally, two angles whose sum is equal to 180 degrees are known as **supplementary angles.** To find the supplement of an angle, subtract the given angle from 180 degrees. For example, the supplement of an angle that is 50 degrees is 180 – 50 = 130 degrees.

These terms involving angles can be seen in many types of word problems. For example, consider the following problem: The measure of an angle is 60 degrees less than two times the measure of its complement. What is the angle's measure? To solve this, let x be the unknown angle. Therefore, its complement is 90 – x. The problem gives that:

$$x = 2(90 - x) - 60$$

To solve for x, distribute the 2, and collect like terms. This process results in:

$$x = 120 - 2x$$

Then, use the addition property to add $2x$ to both sides to obtain $3x = 120$. Finally, use the multiplication properties of equality to divide both sides by 3 to get $x = 40$. Therefore, the angle measures 40 degrees. Also, its complement measures 50 degrees.

Solving for Missing Values in Triangles, Circles, and Other Figures

Solving for missing values in shapes requires knowledge of the shape and its characteristics. For example, a triangle has three sides and three angles that add up to 180 degrees. If two angle measurements are given, the third can be calculated. For the triangle below, the one given angle has a measure of 55 degrees. The missing angle is x. The third angle is labeled with a square, which indicates a measure of 90 degrees. Because all angles must sum to 180 degrees, the following equation can be used to find the missing x-value:

$$55° + 90° + x = 180°$$

Adding the two given angles and subtracting the total from 180, the missing angle is found to be 35 degrees.

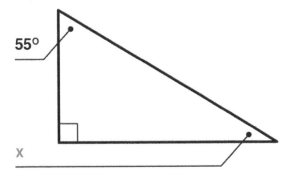

A similar problem can be solved with circles. If the radius is given but the circumference is unknown, the circumference can be calculated based on the formula $C = 2\pi r$. This example can be used in the figure below. The radius can be substituted for r in the formula. Then the circumference can be found as:

$$C = 2\pi \times 8 = 16\pi = 50.24 \text{ cm}$$

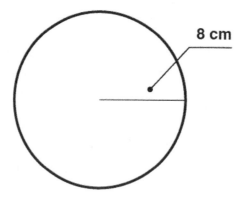

Other figures that may have missing values could be the length of a square, given the area, or the perimeter of a rectangle, given the length and width. All of the missing values can be found by first identifying all the characteristics that are known about the shape, then looking for ways to connect the missing value to the given information.

Measurement

Measuring Lengths of Objects Using Standard Tools

Lengths of objects can be measured using tools such as rulers, yard sticks, meter sticks, and tape measures. Typically, a ruler measures 12 inches, or one foot. For this reason, a ruler is normally used to measure lengths smaller than or just slightly more than 12 inches. Rulers may represent centimeters instead of inches. Some rulers have inches on one side and centimeters on the other. Be sure to recognize what units you are measuring in. The standard ruler measurements are divided into units of 1 inch and normally broken down to $\frac{1}{2}, \frac{1}{4}, \frac{1}{8}$, and even $\frac{1}{16}$ of an inch for more precise measurements. If measuring in centimeters, the centimeter is likely divided into tenths. To measure the size of a picture, for purposes of buying a frame, a ruler is helpful. If the picture is very large, a yardstick, which measures 3 feet and normally is divided into feet and inches, might be useful. Using metric units, the meter stick measures 1 meter and is divided into 100 centimeters. To measure the size of a window in a home, either a yardstick or meter stick would work. To measure the size of a room, though, a tape measure would be the easiest tool to use. Tape measures can measure up to 10 feet, 25 feet, or more depending on the particular tape measure.

Converting Within and Between Standard and Metric Systems

When working with dimensions, sometimes the given units don't match the formula, and conversions must be made. The metric system has base units of meter for length, kilogram for mass, and liter for liquid volume. This system expands to three places above the base unit and three places below. These places correspond with prefixes with a base of 10.

The following table shows the conversions:

kilo-	hecto-	deca-	base	deci-	centi-	milli-
1,000 times the base	100 times the base	10 times the base		1/10 times the base	1/100 times the base	1/1000 times the base

To convert between units within the metric system, values with a base ten can be multiplied. The decimal can also be moved in the direction of the new unit by the same number of zeros on the number. For example, 3 meters is equivalent to 0.003 kilometers. The decimal moved three places (the same number of zeros for kilo-) to the left (the same direction from base to kilo-). Three meters is also equivalent to 3,000 millimeters. The decimal is moved three places to the right because the prefix milli- is three places to the right of the base unit.

The English Standard system used in the United States has a base unit of foot for length, pound for weight, and gallon for liquid volume. These conversions aren't as easy as the metric system because they aren't a base ten model. The following table shows the conversions within this system.

Length	Weight	Capacity
1 foot (ft) = 12 inches (in) 1 yard (yd) = 3 feet 1 mile (mi) = 5280 feet 1 mile = 1760 yards	1 pound (lb) = 16 ounces (oz) 1 ton = 2000 pounds	1 tablespoon (tbsp) = 3 teaspoons (tsp) 1 cup (c) = 16 tablespoons 1 cup = 8 fluid ounces (oz) 1 pint (pt) = 2 cups 1 quart (qt) = 2 pints 1 gallon (gal) = 4 quarts

When converting within the English Standard system, most calculations include a conversion to the base unit and then another to the desired unit. For example, take the following problem: 3 qt = ___ c. There is no straight conversion from quarts to cups, so the first conversion is from quarts to pints. There are 2 pints in 1 quart, so there are 6 pints in 3 quarts. This conversion can be solved as a proportion:

$$\frac{3 \text{ qt}}{x} = \frac{1 \text{ qt}}{2 \text{ pt}}$$

It can also be observed as a ratio 2:1, expanded to 6:3. Then the 6 pints must be converted to cups. The ratio of pints to cups is 1:2, so the expanded ratio is 6:12. For 6 pints, the measurement is 12 cups. This problem can also be set up as one set of fractions to cancel out units. It begins with the given information and cancels out matching units on top and bottom to yield the answer. Consider the following expression:

$$\frac{3 \text{ qt}}{1} \times \frac{2 \text{ pt}}{1 \text{ qt}} \times \frac{2 \text{ c}}{1 \text{ pt}}$$

It's set up so that units on the top and bottom cancel each other out:

$$\frac{3 \text{ q\!t}}{1} \times \frac{2 \text{ p\!t}}{1 \text{ q\!t}} \times \frac{2 \text{ c}}{1 \text{ p\!t}}$$

The numbers can be calculated as 3 × 2 × 2 on the top and 1 on the bottom. It still yields an answer of 12 cups.

This process of setting up fractions and canceling out matching units can be used to convert between standard and metric systems. A few common equivalent conversions are 2.54 cm = 1 in, 3.28 ft = 1 m, and 2.205 lb = 1 kg. Writing these as fractions allows them to be used in conversions. For the fill-in-the-blank problem 5 m = ____ ft, an expression using conversions starts with the expression $\frac{5 \text{ m}}{1} \times \frac{3.28 \text{ ft}}{1 \text{ m}}$, where the units of meters will cancel each other out and the final unit is feet. Calculating the numbers yields 16.4 feet. This problem only required two fractions. Others may require longer expressions, but the underlying rule stays the same. When there's a unit on the top of the fraction that's the same as the unit on the bottom, then they cancel each other out. Using this logic and the conversions given above, many units can be converted between and within the different systems.

The conversion between Fahrenheit and Celsius is found in a formula:

$$°C = (°F - 32) \times \frac{5}{9}$$

For example, to convert 78°F to Celsius, the given temperature would be entered into the formula:

$$°C = (78 - 32) \times \frac{5}{9}$$

Solving the equation, the temperature comes out to be 25.56°C. To convert in the other direction, the formula becomes:

$$°F = °C \times \frac{9}{5} + 32$$

Remember the order of operations when calculating these conversions.

Comparing Relative Sizes of U.S. Customary Units and Metric Units

Measuring length in United States customary units is typically done using inches, feet, yards, and miles. When converting among these units, remember that 12 inches = 1 foot, 3 feet = 1 yard, and 5280 feet = 1 mile. Common customary units of weight are ounces and pounds. The conversion needed is 16 ounces = 1 pound. For customary units of volume ounces, cups, pints, quarts, and gallons are typically used. For conversions, use 8 ounces = 1 cup, 2 cups = 1 pint, 2 pints = 1 quart, and 4 quarts = 1 gallon. For measuring lengths in metric units, know that 100 centimeters = 1 meter, and 1000 meters = 1 kilometer. For metric units of measuring weights, grams and kilograms are often used. Know that 1000 grams = 1 kilogram when making conversions. For metric measures of volume, the most common units are milliliters and liters. Remember that 1000 milliliters = 1 liter.

Data Analysis and Probability

Interpreting Relevant Information from Tables, Charts, and Graphs

Tables, charts, and graphs can be used to convey information about different variables. They are all used to organize, categorize, and compare data, and they all come in different shapes and sizes. Each type has its own way of showing information, whether it is in a column, shape, or picture. To answer a question relating to a table, chart, or graph, some steps should be followed. First, the problem should be read thoroughly to determine what is being asked to determine what quantity is unknown. Then, the title of the table, chart, or graph should be read. The title should clarify what data is actually being summarized in the table. Next, look at the key and labels for both the horizontal and vertical axes, if they are given. These items will provide information about how the data is organized. Finally, look to see if there is any more labeling inside the table. Taking the time to get a good idea of what the table is summarizing will be helpful as it is used to interpret information.

Tables are a good way of showing a lot of information in a small space. The information in a table is organized in columns and rows. For example, a table may be used to show the number of votes each candidate received in an election. By interpreting the table, one may observe which candidate won the election and which candidates came in second and third. In using a bar chart to display monthly rainfall amounts in different countries, rainfall can be compared between countries at different times of the year. Graphs are also a useful way to show change in variables over time, as in a line graph, or percentages of a whole, as in a pie graph.

The table below relates the number of items to the total cost. The table shows that 1 item costs $5. By looking at the table further, 5 items cost $25, 10 items cost $50, and 50 items cost $250. This cost can be extended for any number of items. Since 1 item costs $5, then 2 items would cost $10. Though this information isn't in the table, the given price can be used to calculate unknown information.

Number of Items	1	5	10	50
Cost ($)	5	25	50	250

A **bar graph** is a graph that summarizes data using bars of different heights. It is useful when comparing two or more items or when seeing how a quantity changes over time. It has both a horizontal and vertical axis. Interpreting bar graphs includes recognizing what each bar represents and connecting that to the two variables. The bar graph below shows the scores for six people on three different games. The color of the bar shows which game each person played, and the height of the bar indicates their score for that game. William scored 25 on game 3, and Abigail scored 38 on game 3. By comparing the bars, it's obvious that Williams scored lower than Abigail.

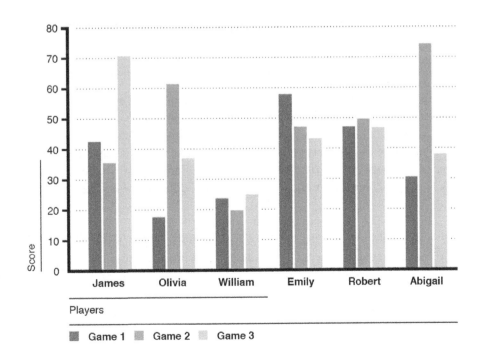

A **line graph** is a way to compare two variables. Each variable is plotted along an axis, and the graph contains both a horizontal and a vertical axis. On a line graph, the line indicates a continuous change. The change can be seen in how the line rises or falls, known as its slope, or rate of change. Often, in line graphs, the horizontal axis represents a variable of time. Audiences can quickly see if an amount has grown or decreased over time. The bottom of the graph, or the x-axis, shows the units for time, such as days, hours, months, etc.

If there are multiple lines, a comparison can be made between what the two lines represent. For example, the following line graph shows the change in temperature over five days. The top line represents the high, and the bottom line represents the low for each day. Looking at the top line alone, the high decreases for a day, then increases on Wednesday. Then it decreased on Thursday and increases again on Friday. The low temperatures have a similar trend, shown in bottom line. The range in temperatures each day can also be calculated by finding the difference between the top line and bottom line on a particular day. On Wednesday, the range was 14 degrees, from 62 to 76° F.

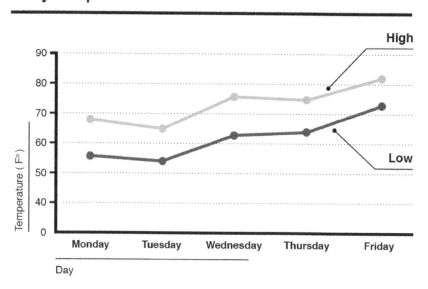

Pie charts are used to show percentages of a whole, as each category is given a piece of the pie, and together all the pieces make up a whole. They are a circular representation of data which are used to highlight numerical proportion. It is true that the arc length of each pie slice is proportional to the amount it individually represents. When a pie chart is shown, an audience can quickly make comparisons by comparing the sizes of the pieces of the pie. They can be useful for comparison between different categories. The following pie chart is a simple example of three different categories shown in comparison to each other.

Light gray represents cats, dark gray represents dogs, and the gray between those two represents other pets. As the pie is cut into three equal pieces, each value represents just more than 33 percent, or $\frac{1}{3}$ of the whole. Values 1 and 2 may be combined to represent $\frac{2}{3}$ of the whole. In an example where the total pie represents 75,000 animals, then cats would be equal to $\frac{1}{3}$ of the total, or 25,000. Dogs would equal 25,000 and other pets also equal 25,000.

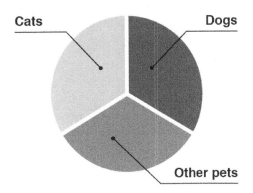

The fact that a circle is 360 degrees is used to create a pie chart. Because each piece of the pie is a percentage of a whole, that percentage is multiplied times 360 to get the number of degrees each piece represents. In the example above, each piece is $\frac{1}{3}$ of the whole, so each piece is equivalent to 120 degrees. Together, all three pieces add up to 360 degrees.

Stacked bar graphs, also used fairly frequently, are used when comparing multiple variables at one time. They combine some elements of both pie charts and bar graphs, using the organization of bar graphs and the proportionality aspect of pie charts. The following is an example of a stacked bar graph that represents the number of students in a band playing drums, flutes, trombones, and clarinets. Each bar graph is broken up further into girls and boys.

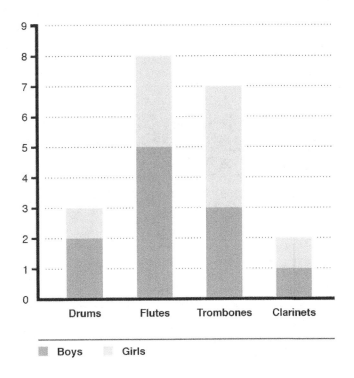

To determine how many boys play tuba, refer to the darker portion of the trombone bar, resulting in 3 students.

A **scatterplot** is another way to represent paired data. It uses Cartesian coordinates, like a line graph, meaning it has both a horizontal and vertical axis. Each data point is represented as a dot on the graph. The dots are never connected with a line. For example, the following is a scatterplot showing people's weight versus height.

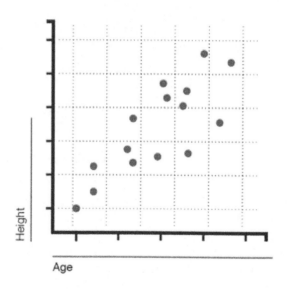

A scatterplot, also known as a **scattergram**, can be used to predict another value and to see if an association, known as a **correlation,** exists between a set of data. If the data resembles a straight line, the data is **associated.** The following is an example of a scatterplot in which the data does not seem to have an association:

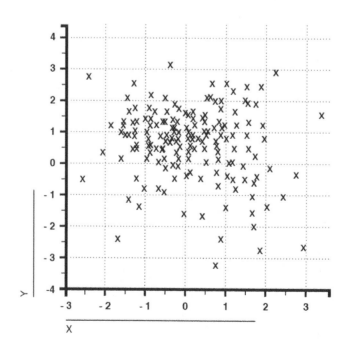

Sets of numbers and other similarly organized data can also be represented graphically. Venn diagrams are a common way to do so. A **Venn diagram** represents each set of data as a circle. The circles overlap, showing that each set of data is overlapping. A Venn diagram is also known as a **logic diagram** because it visualizes all possible logical combinations between two sets. Common elements of two sets are represented by the area of overlap. The following is an example of a Venn diagram of two sets A and B:

Parts of the Venn Diagram

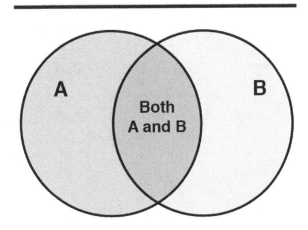

Another name for the area of overlap is the **intersection.** The intersection of A and B, $A \cap B$, contains all elements that are in both sets A and B. The **union** of A and B, $A \cup B$, contains all elements that are in either set A or set B. Finally, the **complement** of $A \cup B$ is equal to all elements that are not in either set A or set B. These elements are placed outside of the circles.

The following is an example of a Venn diagram in which 22 students were surveyed asking about their siblings. Ten students only had a brother, 7 students only had a sister, and 5 had both a brother and a sister. This number 5 is the intersection and is placed where the circles overlap. Two students did not have a brother or a sister. Eight is therefore the complement and is placed outside of the circles.

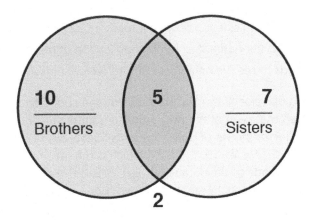

Venn diagrams can have more than two sets of data. The more circles, the more logical combinations are represented by the overlapping. The following is a Venn diagram that represents favorite colors. There were 30 students surveyed. The innermost region represents those students that have a cat, bird, and dog.

Therefore, 2 students had all three. In this example, all students had at least one pet, so no one exists in the complement.

30 students

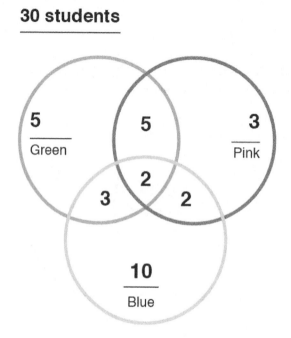

Venn diagrams are typically not drawn to scale, but if they are and their area is proportional to the amount of data it represents, it is known as an **area-proportional Venn diagram**.

Evaluating the Information in Tables, Charts, and Graphs Using Statistics

One way information can be interpreted from tables, charts, and graphs is through statistics. The three most common calculations for a set of data are the mean, median, and mode. These three are called **measures of central tendency**. Measures of central tendency are helpful in comparing two or more different sets of data. The **mean** refers to the average and is found by adding up all values and dividing the total by the number of values. In other words, the mean is equal to the sum of all values divided by the number of data entries. For example, if you bowled a total of 532 points in 4 bowling games, your mean score was $\frac{532}{4} = 133$ points per game. A common application of mean useful to students is calculating what he or she needs to receive on a final exam to receive a desired grade in a class.

The **median** is found by lining up values from least to greatest and choosing the middle value. If there's an even number of values, then the mean of the two middle amounts must be calculated to find the median. For example, the median of the set of dollar amounts $5, $6, $9, $12, and $13 is $9. The **median** of the set of dollar amounts $1, $5, $6, $8, $9, $10 is $7, which is the mean of $6 and $8. The **mode** is the value that occurs the most. The mode of the data set {1, 3, 1, 5, 5, 8, 10} actually refers to two numbers: 1 and 5. In this case, the data set is bimodal because it has two modes. A data set can have no mode if no amount is repeated. Another useful statistic is range. The **range** for a set of data refers to the difference between the highest and lowest value.

In some cases, some numbers in a list of data might have weights attached to them. In that case, a weighted mean can be calculated. A common application of a weighted mean is GPA. In a semester, each class is assigned a number of credit hours, its weight, and at the end of the semester each student receives a grade. To compute GPA, an A is a 4, a B is a 3, a C is a 2, a D is a 1, and an F is a 0. Consider a student that takes a 4-hour English class, a 3-hour math class, and a 4-hour history class and receives all B's. The weighted mean, GPA, is found by multiplying each grade times its weight, number of credit hours, and dividing by the total number of credit hours. Therefore, the student's GPA is:

$$\frac{3 \times 4 + 3 \times 3 + 3 \times 4}{11} = \frac{33}{1} = 3.0.$$

The following bar chart shows how many students attend a cycle on each day of the week. To find the mean attendance for the week, each day's attendance can be added together, $10 + 7 + 6 + 9 + 8 + 14 + 4 = 58$, and the total divided by the number of days, $58 \div 7 = 8.3$. The mean attendance for the week was 8.3 people. The median attendance can be found by putting the attendance numbers in order from least to greatest: 4, 6, 7, 8, 9, 10, 14, and choosing the middle number: 8 people. The mode for attendance is none for this set of data because no numbers repeat. The range is 10, which is found by finding the difference between the lowest number, 4, and the highest number, 14.

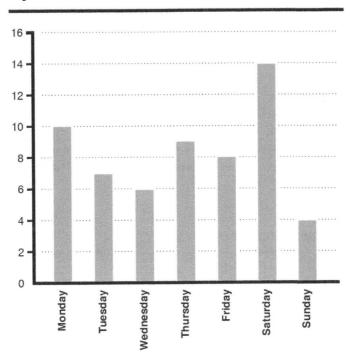

A **histogram** is a bar graph used to group data into "bins" that cover a range on the horizontal, or x-axis. Histograms consist of rectangles whose height is equal to the frequency of a specific category. The horizontal axis represents the specific categories. Because they cover a range of data, these bins have no gaps between bars, unlike the bar graph above. In a histogram showing the heights of adult golden retrievers, the bottom axis would be groups of heights, and the y-axis would be the number of dogs in each range. Evaluating this histogram would show the height of most golden retrievers as falling within a certain range. It also provides information to find the average height and range for how tall golden retrievers may grow.

The following is a histogram that represents exam grades in a given class. The horizontal axis represents ranges of the number of points scored, and the vertical axis represents the number of students. For example, approximately 33 students scored in the 60 to 70 range.

Results of the exam

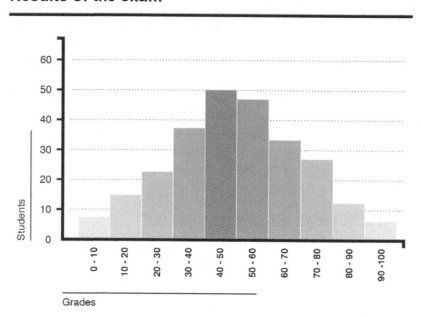

Measures of central tendency can be discussed using a histogram. If the points scored were shown with individual rectangles, the tallest rectangle would represent the mode. A bimodal set of data would have two peaks of equal height. Histograms can be classified as having data **skewed to the left, skewed to the right,** or **normally distributed**, which is also known as **bell-shaped**. These three classifications can be seen in the following chart:

Measures of central tendency images

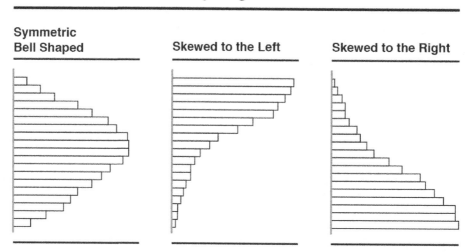

When the data is normal, the mean, median, and mode are all very close. They all represent the most typical value in the data set. The mean is typically used as the best measure of central tendency in this case because it does include all data points. However, if the data is skewed, the mean becomes less meaningful. The median is the best measure of central tendency because it is not affected by any outliers, unlike the mean. When the data is skewed, the mean is dragged in the direction of the skew. Therefore, if the data is not normal, it is best to use the median as the measure of central tendency.

The measures of central tendency and the range may also be found by evaluating information on a line graph.

In the line graph from a previous example that showed the daily high and low temperatures, the average high temperature can be found by gathering data from each day on the triangle line. The days' highs are 82, 78, 75, 65, and 70. The average is found by adding them together to get 370, then dividing by 5 (because there are 5 temperatures). The average high for the five days is 74. If 74 degrees is found on the graph, then it falls in the middle of the values on the triangle line. The mean low temperature can be found in the same way.

Given a set of data, the **correlation coefficient**, r, measures the association between all the data points. If two values are correlated, there is an association between them. However, correlation does not necessarily mean causation, or that one value causes the other. There is a common mistake made that assumes correlation implies causation. Average daily temperature and number of sunbathers are both correlated and have causation. If the temperature increases, that change in weather causes more people to want to catch some rays. However, wearing plus-size clothing and having heart disease are two variables that are correlated but do not have causation. The larger someone is, the more likely he or she is to have heart disease. However, being overweight does not cause someone to have the disease.

Explaining the Relationship between Two Variables

Independent and dependent are two types of variables that describe how they relate to each other. The **independent variable** is the variable controlled by the experimenter. It stands alone and isn't changed by other parts of the experiment. This variable is normally represented by x and is found on the horizontal, or x-axis, of a graph. The **dependent variable** changes in response to the independent variable. It reacts to, or depends on, the independent variable. This variable is normally represented by y and is found on the vertical, or y-axis of the graph.

The relationship between two variables, x and y, can be seen on a scatterplot.

The following scatterplot shows the relationship between weight and height. The graph shows the weight as x and the height as y. The first dot on the left represents a person who is 45 kg and approximately 150 cm tall. The other dots correspond in the same way. As the dots move to the right and weight increases, height also increases. A line could be drawn through the middle of the dots to move from bottom left to top right. This line would indicate a **positive correlation** between the variables. If the variables had a **negative correlation**, then the dots would move from the top left to the bottom right.

Height and Weight

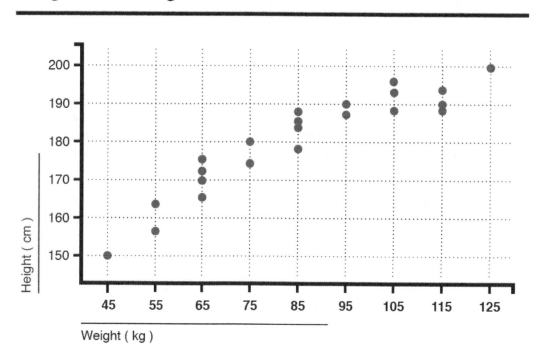

A **scatterplot** is useful in determining the relationship between two variables, but it's not required. Consider an example where a student scores a different grade on his math test for each week of the month. The independent variable would be the weeks of the month. The dependent variable would be the grades, because they change depending on the week. If the grades trended up as the weeks passed, then the relationship between grades and time would be positive. If the grades decreased as the time passed, then the relationship would be negative. (As the number of weeks went up, the grades went down.)

The relationship between two variables can further be described as strong or weak. The relationship between age and height shows a strong positive correlation because children grow taller as they grow up. In adulthood, the relationship between age and height becomes weak, and the dots will spread out. People stop growing in adulthood, and their final heights vary depending on factors like genetics and health. The closer the dots on the graph, the stronger the relationship. As they spread apart, the relationship becomes weaker. If they are too spread out to determine a correlation up or down, then the variables are said to have no correlation.

Variables are values that change, so determining the relationship between them requires an evaluation of who changes them. If the variable changes because of a result in the experiment, then it's dependent. If the variable changes before the experiment, or is changed by the person controlling the experiment, then it's the independent variable. As they interact, one is manipulated by the other. The manipulator is the independent, and the manipulated is the dependent. Once the independent and dependent variable are determined, they can be evaluated to have a positive, negative, or no correlation.

Average and Median

A data set can be described by calculating the mean, median, and mode. These values, called **measures of center,** allow the data to be described with a single value that is representative of the data set.

Again, the most common measure of center is the **mean,** also referred to as the **average.**

To calculate the mean,

- Add all data values together

- Divide by the sample size (the number of data points in the set)

The **median** is middle data value, so that half of the data lies below this value and half lies below the data value.

To calculate the median,

- Order the data from least to greatest

- The point in the middle of the set is the median

    o   In the event that there is an even number of data points, add the two middle points and divide by 2

The **mode** is the data value that occurs most often.

To calculate the mode,

- Order the data from least to greatest

- Find the value that occurs most often

Example: Amelia is a leading scorer on the school's basketball team. The following data set represents the number of points that Amelia has scored in each game this season. Use the mean, median, and mode to describe the data.

16, 12, 26, 14, 28, 14, 12, 15, 25

Solution:

Mean:

$$16 + 12 + 26 + 14 + 28 + 14 + 12 + 15 + 25 = 162$$

$$162 \div 9 = 18$$

Amelia averages 18 points per game.

Median:

12, 12, 14, 14, **15**, 16, 25, 26, 28

Amelia's median score is 15.

Mode:

12, 12, 14, 14, 15, 16, 25, 26, 28

The numbers 12 and 14 each occur twice in the data set, so this set has 2 modes: 12 and 14.

The **range** is the difference between the largest and smallest values in the set. In the example above, the range is 28 – 12 = 16.

Calculating Probabilities

Probability describes how likely it is that an event will occur. Probabilities are always a number from zero to 1. If an event has a high likelihood of occurrence, it will have a probability close to 1. If there is only a small chance that an event will occur, the likelihood is close to zero. A fair six-sided die has one of the numbers 1, 2, 3, 4, 5, and 6 on each side. When this die is rolled there is a one in six chance that it will land on 2. This is because there are six possibilities and only one side has a 2 on it. The probability then is $\frac{1}{6}$ or 0.167. The probability of rolling an even number from this die is three in six, which is $\frac{1}{2}$ or 0.5. This is because there are three sides on the die with even numbers (2, 4, 6), and there are six possible sides. The probability of rolling a number less than 10 is one because every side of the die has a number less than 6, so this is certain to occur. On the other hand, the probability of rolling a number larger than 20 is zero. There are no numbers greater than 20 on the die, so it is certain that this will not occur, thus the probability is zero.

If a teacher says that the probability of anyone passing her final exam is 0.2, is it highly likely that anyone will pass? No, the probability of anyone passing her exam is low because 0.2 is closer to zero than to 1. If another teacher is proud that the probability of students passing his class is 0.95, how likely is it that a student will pass? It is highly likely that a student will pass because the probability, 0.95, is very close to 1.

Expressing Probabilities in a Variety of Ways

Probabilities can be expressed in a variety of ways, such as ratios, proportions, decimals, and percents. For example, consider a coin purse that contains ten coins: 3 pennies, 2 nickels, 4 dimes, and 1 quarter. If every coin has an equal chance of getting drawn from the purse, the likelihood of getting a nickel is 2/10, which reduces to 1/5. This fraction can be written as the decimal 0.2, the percent 20%, and the proportion 2:10. Regardless as to the representation format, each of these indicate the same likelihood of selecting a nickel from the set of 10 coins.

Problem Solving

Solving Problems Involving Elapsed Time, Money, Length, Volume, and Mass

To solve problems, follow these steps: Identify the variables that are known, decide which equation should be used, substitute the numbers, and solve. To solve an equation for the amount of time that has elapsed since an event, use the equation T = L – E where T represents the elapsed time, L represents the later time, and E represents the earlier time. For example, the Minnesota Vikings have not appeared in the Super Bowl since 1976. If the year is now 2017, how long has it been since the Vikings were in the Super Bowl? The later time, L, is 2017, E = 1976, and the unknown is T. Substituting these numbers, the equation is T = 2017 – 1976, and so T = 41. It has been 41 years since the Vikings have appeared in the

Super Bowl. Questions involving total cost can be solved using the formula, $C = I + T$ where C represents the total cost, I represents the cost of the item purchased, and T represents the tax amount. To find the length of a rectangle given the area = 32 square inches and width = 8 inches, the formula $A = L \times W$ can be used. Substitute 32 for A and substitute 8 for w, giving the equation $32 = L \times 8$. This equation is solved by dividing both sides by 8 to find that the length of the rectangle is 4. The formula for volume of a rectangular prism is given by the equation $V = L \times W \times H$. If the length of a rectangular juice box is 4 centimeters, the width is 2 centimeters, and the height is 8 centimeters, what is the volume of this box? Substituting in the formula we find $V = 4 \times 2 \times 8$, so the volume is 64 cubic centimeters. In a similar fashion as those previously shown, the mass of an object can be calculated given the formula, Mass = Density × Volume.

Using Inverse Operations to Solve Problems

Inverse operations can be used to solve problems where there is a missing value. The area for a rectangle may be given, along with the length, but the width may be unknown. This situation can be modeled by the equation Area = Length × Width. The area is 40 square feet and the length is 10 feet. The equation becomes $40 = 10 \times w$. In order to find the w, we recognize that some number multiplied by 10 yields the number 40. The inverse operation to multiplication is division, so the 10 can be divided on both sides of the equation. This operation cancels out the 10 and yields an answer of 4 for the width. The following equation shows the work:

$$40 = 10 \times w$$

$$\frac{40}{10} = \frac{10 \times w}{10}$$

$$4 = w$$

Other inverse operations can be used to solve problems as well. The following equation can be solved for b: $b + 4 = 9$. Because 4 is added to b, it can be subtracted on both sides of the equal sign to cancel out the four and solve for b, as follows:

$$b + 4 - 4 = 9 - 4$$

$$b = 5$$

Whatever operation is used in the equation, the inverse operation can be used and applied to both sides of the equals sign to solve for an unknown value.

Quantitative Reasoning Practice Questions

1. Joey and Sandy wanted to sell lemonade and cookies to earn some extra money. They sold cookies for $1 and lemonade for $0.50. At the end of the first day they had sold 24 cookies. The total money that they collected was $35.50. How many cups of lemonade did they sell?

    a. 11

    b. 21

    c. 22

    d. 23

2. Carla is starting a cake-decorating business and wants to know how long it will take her to start making a profit. She knows the original investment is $100. After that investment, she can begin making cakes and selling them for $20 each. How many cakes will she need to sell to break even on her investment?

    a. 5

    b. 100

    c. 10

    d. 2

3. The sides of a triangle have the following lengths: 4 inches, 4 inches, and 7.5 inches. If the smallest angle within the triangle has a measurement of 20°, what is the measure of the largest angle?

    a. 100°

    b. 140°

    c. 120°

    d. 160°

4. On the first four tests this semester, a student received the following scores out of 100: 74, 76, 82, and 84. The student must earn at least what score on the fifth test to receive a B in the class? Assume that the final test is also out of 100 points and that to receive a B in the class, he must have at least an 80% average.

    a. 80

    b. 84

    c. 82

    d. 78

5. A jar is filled with green, yellow, and orange marbles. If $\frac{1}{4}$ of the marbles are green and $\frac{2}{7}$ are yellow, what fraction of the marbles are orange?

    a. $\frac{15}{28}$

    b. $\frac{13}{28}$

    c. $\frac{2}{3}$

    d. $\frac{3}{7}$

Quantitative Reasoning Answer Explanations

1. D: The following equation can be used to model how much money they collect for their sale: $M = 1c + 0.5l$, where M is money collected and c is cookies and l is lemonade. By substituting 35.50 in for the variable M and 24 for the variable c, the equation can be solved for l, which is found to be 23 cups of lemonade. If there were 11 lemonades sold, the money collected would be \$29.50. If 21 lemonades were sold, the money collected would be \$34.50. If 22 lemonades were sold, the money collected would be \$35.

2. A: The equation used to model this situation is $y = 20x - 100$, where 20 is price of each cake and 100 is the original investment. The value of x is the number of cakes and y is the money she makes. If this line is plotted on the graph, the x-intercept will be the number of cakes she needs to make in order to recoup her investment and break even. The x-intercept occurs when the y-value is zero. For this equation, setting $y = 0$ and solving for x gives a value of 5 cakes. 100 cakes would yield a profit of \$1,900, ten cakes would yield a profit of \$100, and 2 cakes would still leave her in the negative (-\$10).

3. B: Because two sides are equal, this is an isosceles triangle. The smallest sides correspond to the 20° angles. The third angle has a measure of:

$$180 - 20 - 20 = 140°$$

4. B: Let x be equal to the fifth test score. Therefore, in order to receive, at minimum, a B in the class, the student must have:

$$\frac{74 + 76 + 82 + 84 + x}{5} = 80$$

Therefore:

$$\frac{316 + x}{5} = 80$$

Solving for x gives $316 + x = 400$, or $x = 84$. Therefore, he must receive at least an 84 out of 100 on the fifth test to receive a B in the course.

5. B: The total fraction of green and yellow marbles is:

$$\frac{1}{4} + \frac{2}{7} = \frac{7}{28} + \frac{8}{28} = \frac{15}{28}$$

Mathematics Achievement Practice Questions

1. How many centimeters are in 3 feet? (Note: 2.54 cm = 1 in)
   a. 0.635
   b. 91.44
   c. 14.17
   d. 7.62

2. Which of the following is the equation of a vertical line that runs through the point (1, 4)?
   a. $x = 1$
   b. $y = 1$
   c. $x = 4$
   d. $y = 4$

3. If the ratio of x to y is 1:8, what is the product of x and y when $y = 48$?
   a. 96
   b. 183
   c. 48
   d. 288

4. The percent increase from 8 to 18 is equivalent to the percent increase from 234 to what number?
   a. 468
   b. 526
   c. 526.5
   d. 125

5. What is the perimeter of the following figure?

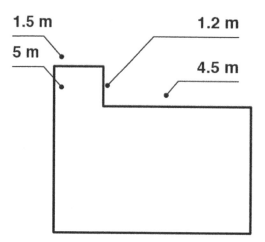

   a. 13.4 m
   b. 22 m
   c. 12.2 m
   d. 22.5 m

Mathematics Achievement Answer Explanations

1. B: The conversion between feet and centimeters requires a middle term. As there are 2.54 centimeters in 1 inch, the conversion between inches and feet must be found. As there are 12 inches in a foot, the fractions can be set up as follows:

$$3 \text{ ft} \times \frac{12 \text{ in}}{1 \text{ ft}} \times \frac{2.54 \text{ cm}}{1 \text{ in}}$$

The feet and inches cancel out to leave only centimeters for the answer. The numbers are calculated across the top and bottom to yield:

$$\frac{3 \times 12 \times 2.54}{1 \times 1} = 91.44$$

The number and units used together form the answer of 91.44 cm.

2. A: A vertical line has the same x value for any point on the line. Other points on the line would be (1, 3), (1, 5), (1, 9), etc. Mathematically, this is written as $x = 1$. A vertical line is always of the form $x = a$ for some constant a.

3. D: The ratio gives the proportion $\frac{x}{y} = \frac{1}{8}$. If $y = 48$, then $\frac{x}{48} = \frac{1}{8}$ means that $x = 6$. The product of x and y is therefore (6)(48) = 288.

4. C: First, calculate the percent increase from 8 to 18 as:

$$\frac{18 - 8}{8} = 1.25 = 125\%$$

Then add 125% of 234 onto 234 to obtain:

$$292.5 + 234 = 526.5$$

5. B: The perimeter is found by adding the length of all the exterior sides. When the given dimensions are added, the perimeter is 22 meters. The equation to find the perimeter can be:

$$P = 5 + 1.5 + 1.2 + 4.5 + 3.8 + 6 = 22$$

The last two dimensions can be found by subtracting 1.2 from 5, and adding 1.5 and 4.5, respectively.

Reading Comprehension

Identifying the Topic, Main Idea, and Supporting Details

The **topic** of a text is the general subject matter. Text topics can usually be expressed in one word, or a few words at most. Additionally, readers should ask themselves what point the author is trying to make. This point is the **main idea** of the text, the one thing the author wants readers to know concerning the topic. Once the author has established the main idea, they will support the main idea by supporting details. **Supporting details** are evidence that support the main idea and include personal testimonies, examples, or statistics.

One analogy for these components and their relationships is that a text is like a well-designed house. The topic is the roof, covering all rooms. The main idea is the frame. The supporting details are the various rooms. To identify the topic of a text, readers can ask themselves what or who the author is writing about in the paragraph. To locate the main idea, readers can ask themselves what one idea the author wants readers to know about the topic. To identify supporting details, readers can put the main idea into question form and ask, "what does the author use to prove or explain their main idea?"

Let's look at an example. An author is writing an essay about the Amazon rainforest and trying to convince the audience that more funding should go into protecting the area from deforestation. The author makes the argument stronger by including evidence of the benefits of the rainforest: it provides habitats to a variety of species, it provides much of the earth's oxygen which in turn cleans the atmosphere, and it is the home to medicinal plants that may be the answer to some of the world's deadliest diseases.

Here is an outline of the essay looking at topic, main idea, and supporting details:

Topic: Amazon rainforest
Main Idea: The Amazon rainforest should receive more funding in order to protect it from deforestation.
Supporting Details:
    1. It provides habitats to a variety of species
    2. It provides much of the earth's oxygen which in turn cleans the atmosphere
    3. It is home to medicinal plants that may be the answer to some of the world's deadliest diseases.

Notice that the topic of the essay is listed in a few key words: "Amazon rainforest." The main idea tells us what about the topic is important: that the topic should be funded in order to prevent deforestation. Finally, the supporting details are what author relies on to convince the audience to act or to believe in the truth of the main idea.

Inferences

Making an inference from a selection means to make an educated guess from the passage read. **Inferences** should be conclusions based off of sound evidence and reasoning. When multiple-choice test questions ask about the logical conclusion that can be drawn from reading text, the test-taker must identify which choice will unavoidably lead to that conclusion. In order to eliminate the incorrect choices, the test-taker should come up with a hypothetical situation wherein an answer choice is true, but the conclusion is not true.

For example, here is an example with three answer choices:

Fred purchased the newest PC available on the market. Therefore, he purchased the most expensive PC in the computer store.

What can one assume for this conclusion to follow logically?

a. Fred enjoys purchasing expensive items.
b. PCs are some of the most expensive personal technology products available.
c. The newest PC is the most expensive one.

The premise of the text is the first sentence: Fred purchased the newest PC. The conclusion is the second sentence: Fred purchased the most expensive PC. Recent release and price are two different factors; the difference between them is the logical gap. To eliminate the gap, one must equate whatever new information the conclusion introduces with the pertinent information the premise has stated. This example simplifies the process by having only one of each: one must equate product recency with product price. Therefore, a possible bridge to the logical gap could be a sentence stating that the newest PCs always cost the most.

Organization/Logic

Recognizing the Structure of Texts in Various Formats
Text structure is the way in which the author organizes and presents textual information so readers can follow and comprehend it. One kind of text structure is sequence. This means the author arranges the text in a logical order from beginning to middle to end. There are three types of sequences:

- Chronological: ordering events in time from earliest to latest

- Spatial: describing objects, people, or spaces according to their relationships to one another in space

- Order of Importance: addressing topics, characters, or ideas according to how important they are, from either least important to most important

Chronological sequence is the most common sequential text structure. Readers can identify sequential structure by looking for words that signal it, like *first, earlier, meanwhile, next, then, later, finally;* and specific times and dates the author includes as chronological references.

Problem-Solution Text Structure
The problem-solution text structure organizes textual information by presenting readers with a problem and then developing its solution throughout the course of the text. The author may present a variety of alternatives as possible solutions, eliminating each as they are found unsuccessful, or gradually leading up to the ultimate solution. For example, in fiction, an author might write a murder mystery novel and have the character(s) solve it through investigating various clues or character alibis until the killer is identified. In nonfiction, an author writing an essay or book on a real-world problem might discuss various alternatives and explain their disadvantages or why they would not work before identifying the best solution. For scientific research, an author reporting and discussing scientific experiment results would explain why various alternatives failed or succeeded.

Comparison-Contrast Text Structure

Comparison identifies similarities between two or more things. **Contrast** identifies differences between two or more things. Authors typically employ both to illustrate relationships between things by highlighting their commonalities and deviations. For example, a writer might compare Windows and Linux as operating systems, and contrast Linux as free and open-source vs. Windows as proprietary. When writing an essay, sometimes it is useful to create an image of the two objects or events you are comparing or contrasting. Venn diagrams are useful because they show the differences as well as the similarities between two things. Once you've seen the similarities and differences on paper, it might be helpful to create an outline of the essay with both comparison and contrast. Every outline will look different, because every two or more things will have a different number of comparisons and contrasts. Say you are trying to compare and contrast carrots with sweet potatoes.

Here is an example of a compare/contrast outline using those topics:

- Introduction: Talk about why you are comparing and contrasting carrots and sweet potatoes. Give the thesis statement.
- Body paragraph 1: Sweet potatoes and carrots are both root vegetables (similarity)
- Body paragraph 2: Sweet potatoes and carrots are both orange (similarity)
- Body paragraph 3: Sweet potatoes and carrots have different nutritional components (difference)
- Conclusion: Restate the purpose of your comparison/contrast essay.

Of course, if there is only one similarity between your topics and two differences, you will want to rearrange your outline. Always tailor your essay to what works best with your topic.

Descriptive Text Structure

Description can be both a type of text structure and a type of text. Some texts are descriptive throughout entire books. For example, a book may describe the geography of a certain country, state, or region, or tell readers all about dolphins by describing many of their characteristics. Many other texts are not descriptive throughout, but use descriptive passages within the overall text. The following are a few examples of descriptive text:

- When the author describes a character in a novel
- When the author sets the scene for an event by describing the setting
- When a biographer describes the personality and behaviors of a real-life individual
- When a historian describes the details of a particular battle within a book about a specific war
- When a travel writer describes the climate, people, foods, and/or customs of a certain place

A hallmark of description is using sensory details, painting a vivid picture so readers can imagine it almost as if they were experiencing it personally.

Cause and Effect Text Structure

When using cause and effect to extrapolate meaning from text, readers must determine the cause when the author only communicates effects. For example, if a description of a child eating an ice cream cone includes details like beads of sweat forming on the child's face and the ice cream dripping down her hand faster than she can lick it off, the reader can infer or conclude it must be hot outside. A useful technique for making such decisions is wording them in "*If...then*" form, e.g. "*If* the child is perspiring and the ice cream melting, *then* it may be a hot day." Cause and effect text structures explain why certain events or actions resulted in particular outcomes. For example, an author might describe America's historical large flocks of dodo birds, the fact that gunshots did not startle/frighten dodos, and that because dodos did not flee, settlers killed whole flocks in one hunting session, explaining how the dodo was hunted into extinction.

Summarizing a Complex Text

An important skill is the ability to read a complex text and then reduce its length and complexity by focusing on the key events and details. A **summary** is a shortened version of the original text, written by the reader in their own words. The summary should be shorter than the original text, and it must be thoughtfully formed to include critical points from the original text.

In order to effectively summarize a complex text, it's necessary to understand the original source and identify the major points covered. It may be helpful to outline the original text to get the big picture and avoid getting bogged down in the minor details. For example, a summary wouldn't include a statistic from the original source unless it was the major focus of the text. It's also important for readers to use their own words, yet retain the original meaning of the passage. The key to a good summary is emphasizing the main idea without changing the focus of the original information.

The more complex a text, the more difficult it can be to summarize. Readers must evaluate all points from the original source and then filter out what they feel are the less necessary details. Only the essential ideas should remain. The summary often mirrors the original text's organizational structure. For example, in a problem-solution text structure, the author typically presents readers with a problem and then develops solutions through the course of the text. An effective summary would likely retain this general structure, rephrasing the problem and then reporting the most useful or plausible solutions.

Paraphrasing is somewhat similar to summarizing. It calls for the reader to take a small part of the passage and list or describe its main points. Paraphrasing is more than rewording the original passage, though. As with summary, a paraphrase should be written in the reader's own words, while still retaining the meaning of the original source. The main difference between summarizing and paraphrasing is that a summary would be appropriate for a much larger text, while paraphrase might focus on just a few lines of text. Effective paraphrasing will indicate an understanding of the original source, yet still help the reader expand on their interpretation. A paraphrase should neither add new information nor remove essential facts that change the meaning of the source.

Tone and Style

Some question stems will ask about the author's attitude toward a certain person or idea. While it may seem impossible to know exactly what the author felt toward their subject, there are clues to indicate the emotion, or lack thereof, of the author. Clues like word choice or style will alert readers to the author's attitude. Some possible words that name the author's attitude are listed below:

- Admiring
- Angry
- Critical
- Defensive
- Enthusiastic
- Humorous
- Moralizing
- Neutral
- Objective
- Patriotic
- Persuasive

- Playful
- Sentimental
- Serious
- Supportive
- Sympathetic
- Unsupportive

An author's tone is the author's attitude toward their subject and is usually indicated by word choice. If an author's attitude toward their subject is one of disdain, the author will show the subject in a negative light, using deflating words or words that are negatively-charged. If an author's attitude toward their subject is one of praise, the author will use agreeable words and show the subject in a positive light. If an author takes a neutral tone towards their subject, their words will be neutral as well, and they probably will show all sides of their subject, not just the negative or positive side.

Style is another indication of the author's attitude and includes aspects such as sentence structure, type of language, and formatting. Sentence structure is how a sentence is put together. Sometimes, short, choppy sentences will indicate a certain tone given the surrounding context, while longer sentences may serve to create a buffer to avoid being too harsh, or may be used to explain additional information. Style may also include formal or informal language. Using formal language to talk about a subject may indicate a level of respect. Using informal language may be used to create an atmosphere of friendliness or familiarity with a subject. Again, it depends on the surrounding context whether or not language is used in a negative or positive way. Style may also include formatting, such as determining the length of paragraphs or figuring out how to address the reader at the very beginning of the text.

Figurative Language

Authors of a text use language with multiple levels of meaning for many different reasons. When the meaning of a text calls for directness, literal language should be used to provide clarity to the reader. Figurative language can be used when the author wants to produce an emotional effect in the reader or facilitate a deeper understanding of a word or passage. For example, if someone wanted to write a set of instructions on how to use a computer, they would write in literal language. However, if someone wanted to comment on the social implications of banning immigration, they might want to use a wide range of figurative language to highlight an empathetic response. It is important to keep in mind, too, that a single text can have a mixture of both literal and figurative language.

Literal Language
Literal language uses words in accordance with their actual definition. Many informational texts employ literal language because it is straightforward and precise. Documents such as instructions, proposals, technical documents, and workplace documents use literal language for the majority of their writing, so there is no confusion or complexity of meaning for readers to decipher. The information is best communicated through clear and precise language. The following are brief examples of literal language:

- I cook with olive oil.
- There are 365 days in a year.
- My grandma's name is Barbara.
- Yesterday we had some scattered thunderstorms.
- World War II began in 1939.
- Blue whales are the largest species of whale.

Figurative Language

Not meant to be taken literal, figurative language is useful when the author of a text wants to produce an emotional effect in the reader or add a heightened complexity to the meaning of the text. Figurative language is used more heavily in texts such as literary fiction, poetry, critical theory, and speeches. Figurative language goes beyond literal language, allowing readers to form associations they wouldn't normally form with literal language. Using language in a figurative sense appeals to the imagination of the reader. It is important to remember that words themselves are signifiers of objects and ideas, and not the objects and ideas themselves. Figurative language can highlight this detachment by creating multiple associations, but also points to the fact that language is fluid and capable of creating a world full of linguistic possibilities. Figurative language, it can be argued, is the heart of communication even outside of fiction and poetry. People connect through humor, metaphors, cultural allusions, puns, and symbolism in their everyday rhetoric. The following are terms associated with figurative language:

Simile

A simile is a comparison of two things using *like*, *than*, or *as*. A simile usually takes objects that have no apparent connection, such as a mind and an orchid, and compares them:

> His mind was as complex and rare as a field of ghost orchids.

Similes encourage a new, fresh perspective on objects or ideas that wouldn't otherwise occur. Similes are different than metaphors. Metaphors do not use *like*, *than*, or *as*. So, a metaphor from the above example would be:

> His mind was a field of ghost orchids.

Thus, similes highlight the comparison by focusing on the figurative side of the language, elucidating more the author's intent: a field of ghost orchids is something complex and rare, like the mind of a genius. With the metaphor, however, we get a beautiful yet somewhat equivocal comparison.

Metaphor

A popular use of figurative language, metaphors compare objects or ideas directly, asserting that something *is* a certain thing, even if it isn't. The following is an example of a metaphor used by writer Virginia Woolf:

> Books are the mirrors of the soul.

Metaphors have a vehicle and a tenor. The tenor is "books" and the vehicle is "mirrors of the soul." That is, the tenor is what is meant to be described, and the vehicle is that which carries the weight of the comparison. In this metaphor, perhaps the author means to say that written language (books) reflect a person's most inner thoughts and desires.

There are also dead metaphors, which means that the phrases have been so overused to the point where the figurative meaning becomes literal, like the phrase "What you're saying is crystal clear." The phrase compares "what's being said" to something "crystal clear." However, since the latter part of the phrase is in such popular use, the meaning seems literal ("I understand what you're saying") even when it's not.

Finally, an extended metaphor is a metaphor that goes on for several paragraphs, or even an entire text. John Keats' poem "On First Looking into Chapman's Homer" begins, "Much have I travell'd in the realms of gold," and goes on to explain the first time he hears Chapman's translation of Homer's writing. We see the extended metaphor begin in the first line. Keats is comparing travelling into "realms of gold" and exploration of new lands to the act of hearing a certain kind of literature for the first time. The extended metaphor goes on until the end of the poem where Keats stands "Silent, upon a peak in Darien," having heard the end of Chapman's translation. Keats has gained insight into new lands (new text) and is the richer for it.

The following are brief definitions and examples of popular figurative language:

Onomatopoeia: A word that, when spoken, imitates the sound to which it refers. Ex: "We heard a loud *boom* while driving to the beach yesterday."

Personification: When human characteristics are given to animals, inanimate objects, or abstractions. An example would be in William Wordsworth's poem "Daffodils" where he sees a "crowd . . . / of golden daffodils . . . / Fluttering and dancing in the breeze." Dancing is usually a characteristic attributed solely to humans, but Wordsworth personifies the daffodils here as a crowd of people dancing.

Juxtaposition: Juxtaposition is placing two objects side by side for comparison. In literature, this might look like placing two characters side by side for contrasting effect, like God and Satan in Milton's "Paradise Lost."

Paradox: A paradox is a statement that is self-contradictory but will be found nonetheless true. One example of a paradoxical phrase is when Socrates said, "I know one thing; that I know nothing." Seemingly, if Socrates knew nothing, he wouldn't know that he knew nothing. However, it is one thing he knows: that true wisdom begins with casting all presuppositions one has about the world aside.

Hyperbole: A hyperbole is an exaggeration. Ex: "I'm so tired I could sleep for centuries."

Allusion: An allusion is a reference to a character or event that happened in the past. An example of a poem littered with allusions is T.S. Eliot's "The Waste Land." An example of a biblical allusion manifests when the poet says, "I will show you fear in a handful of dust," creating an ominous tone from Genesis 3:19 "For you are dust, and to dust you shall return."

Pun: Puns are used in popular culture to invoke humor by exploiting the meanings of words. They can also be used in literature to give hints of meaning in unexpected places. One example of a pun is when Mercutio is giving his monologue after he is stabbed by Tybalt in "Romeo and Juliet" and says, "look for me tomorrow and you will find me a grave man."

Imagery: This is a collection of images given to the reader by the author. If a text is rich in imagery, it is easier for the reader to imagine themselves in the author's world. One example of a poem that relies on imagery is William Carlos Williams' "The Red Wheelbarrow":

> so much depends
> upon
>
> a red wheel
> barrow
>
> glazed with rain
> water
>
> beside the white
> chickens

The starkness of the imagery and the placement of the words in the poem, to some readers, throw the poem into a meditative state where, indeed, the world of this poem is made up solely of images of a purely simple life. This poem tells a story in sixteen words by using imagery.

Symbolism: A symbol is used to represent an idea or belief system. For example, poets in Western civilization have been using the symbol of a rose for hundreds of years to represent love. In Japan, poets have used the firefly to symbolize passionate love, and sometimes even spirits of those who have died. Symbols can also express powerful political commentary and can be used in propaganda.

Irony: There are three types of irony. Verbal irony is when a person states one thing and means the opposite. For example, a person is probably using irony when they say, "I can't wait to study for this exam next week." Dramatic irony occurs in a narrative and happens when the audience knows something that the characters do not. In the modern TV series *Hannibal*, we as an audience know that Hannibal Lecter is a serial killer, but most of the main characters do not. This is dramatic irony. Finally, situational irony is when one expects something to happen, and the opposite occurs. For example, we can say that a fire station burning down would be an instance of situational irony.

Practice Questions

The following three questions are based on the book On the Trail *by Lina Beard and Adelia Belle Beard.*

For any journey, by rail or by boat, one has a general idea of the direction to be taken, the character of the land or water to be crossed, and of what one will find at the end. So it should be in striking the trail. Learn all you can about the path you are to follow. Whether it is plain or obscure, wet or dry; where it leads; and its length, measured more by time than by actual miles. A smooth, even trail of five miles will not consume the time and strength that must be expended upon a trail of half that length which leads over uneven ground, varied by bogs and obstructed by rocks and fallen trees, or a trail that is all up-hill climbing. If you are a novice and accustomed to walking only over smooth and level ground, you must allow more time for covering the distance than an experienced person would require and must count upon the expenditure of more strength, because your feet are not trained to the wilderness paths with their pitfalls and traps for the unwary, and every nerve and muscle will be strained to secure a safe foothold amid the tangled roots, on the slippery, moss-covered logs, over precipitous rocks that lie in your path. It will take time to pick your way over boggy places where the water oozes up through the thin, loamy soil as through a sponge; and experience alone will teach you which hummock of grass or moss will make a safe stepping-place and will not sink beneath your weight and soak your feet with hidden water. Do not scorn to learn all you can about the trail you are to take . . . It is not that you hesitate to encounter difficulties, but that you may prepare for them. In unknown regions take a responsible guide with you, unless the trail is short, easily followed, and a frequented one. Do not go alone through lonely places; and, being on the trail, keep it and try no explorations of your own, at least not until you are quite familiar with the country and the ways of the wild.

1. What does the author say about unknown regions?
    a. You should try and explore unknown regions in order to learn the land better.
    b. Unless the trail is short or frequented, you should take a responsible guide with you.
    c. All unknown regions will contain pitfalls, traps, and boggy places.
    d. It's better to travel unknown regions by rail rather than by foot.

2. Which of the following is NOT a detail from the passage.
    a. Learning about the trail beforehand is imperative
    b. Time will differ depending on the land
    c. Once you are familiar with the outdoors you can go places on your own
    d. Be careful for wild animals on the trail you are on

3. The passage above is considered which of the following types of writing?
    a. Descriptive
    b. Persuasive
    c. Narrative
    d. Informative

The next two questions are based on this passage, which taken from Chapter 6 of Sense and Sensibility, *by Jane Austen:*

> As a house, Barton Cottage, though small, was comfortable and compact; but as a cottage it was defective, for the building was regular, the roof was tiled, the window shutters were not painted green, nor were the walls covered with honeysuckles. A narrow passage led directly through the house into the garden behind. On each side of the entrance was a sitting room, about sixteen feet square; and beyond them were the offices and the stairs. Four bed-rooms and two garrets formed the rest of the house. It had not been built many years and was in good repair. In comparison of Norland, it was poor and small indeed!—but the tears which recollection called forth as they entered the house were soon dried away. They were cheered by the joy of the servants on their arrival, and each for the sake of the others resolved to appear happy. It was very early in September; the season was fine, and from first seeing the place under the advantage of good weather, they received an impression in its favour which was of material service in recommending it to their lasting approbation.

4. Which of the following is this passage describing?
    a. Museum
    b. Cottage
    c. Skyscraper
    d. Island

5. The narrow passage led through the house into which of the following?
    a. Office
    b. Bedroom
    c. Kitchen
    d. Garden

Answer Explanations

1. B: Choice *B* is the best answer here; the sentence states "In unknown regions take a responsible guide with you, unless the trail is short, easily followed, and a frequented one." Choice *A* is incorrect; the passage does not state that you should try and explore unknown regions. Choice *C* is incorrect; the passage talks about trails that contain pitfalls, traps, and boggy places, but it does not say that *all* unknown regions contain these things. Choice *D* is incorrect; the passage mentions "rail" and "boat" as means of transport at the beginning, but it does not suggest it is better to travel unknown regions by rail.

2. D: Choice *D* is correct; it may be real advice an experienced hiker would give to an inexperienced hiker. However, the question asks about details in the passage, and this is not in the passage. Choice *A* is incorrect; we do see the author encouraging the reader to learn about the trail beforehand . . . "wet or dry; where it leads; and its length." Choice *B* is also incorrect, because we do see the author telling us the time will lengthen with boggy or rugged places opposed to smooth places. Choice *C* is incorrect; at the end of the passage, the author tells us "do not go alone through lonely places . . . unless you are quite familiar with the country and the ways of the wild."

3. D: This is an informative passage. Informative passages explain to the readers how to do something; in this case, the author is attempting to explain the fundamentals of camping and hiking. Descriptive is a type of passage describing a character, event, or place in great detail and imagery, so this is incorrect. A persuasive passage is an argument that tries to get readers to agree with something. A narrative is a passage that tells a story, so this is also incorrect.

4. B: The passage is describing a cottage, Choice *B*. The name of the building is called "Barton Cottage."

5. D: The middle of the passage says that the narrow passage led through the house into the garden, Choice *D*.

Essay

Elements of the Writing Process

Skilled writers undergo a series of steps that comprise the writing process. The purpose of adhering to a structured approach to writing is to develop clear, meaningful, coherent work.

The stages are pre-writing or planning, organizing, drafting/writing, revising, and editing. Not every writer will necessarily follow all five stages for every project but will judiciously employ the crucial components of the stages for most formal or important work. For example, a brief informal response to a short reading passage may not necessitate the need for significant organization after idea generation, but larger assignments and essays will likely mandate use of the full process.

Pre-Writing/Planning

Brainstorming

One of the most important steps in writing is pre-writing. Before drafting an essay or other assignment, it's helpful to think about the topic for a moment or two, in order to gain a more solid understanding of what the task is. Then, spend about five minutes jotting down the immediate ideas that could work for the essay. **Brainstorming** is a way to get some words on the page and offer a reference for ideas when drafting. Scratch paper is provided for writers to use any pre-writing techniques such as webbing, freewriting, or listing. Some writers prefer using graphic organizers during this phase. The goal is to get ideas out of the mind and onto the page.

Freewriting

Like brainstorming, **freewriting** is another prewriting activity to help the writer generate ideas. This method involves setting a timer for two or three minutes and writing down all ideas that come to mind about the topic using complete sentences. Once time is up, writers should review the sentences to see what observations have been made and how these ideas might translate into a more unified direction for the topic. Even if sentences lack sense as a whole, freewriting is an excellent way to get ideas onto the page in the very beginning stages of writing. Using complete sentences can make this a bit more challenging than brainstorming, but overall it is a worthwhile exercise, as it may force the writer to come up with more complete thoughts about the topic.

Once the ideas are on the page, it's time for the writer to turn them into a solid plan for the essay. The best ideas from the brainstorming results can then be developed into a more formal outline.

Organizing

Although sometimes it is difficult to get going on the brainstorming or prewriting phase, once ideas start flowing, writers often find that they have amassed too many thoughts that will not make for a cohesive and unified essay. During the organization stage, writers should examine the generated ideas, hone in on the important ones central to their main idea, and arrange the points in a logical and effective manner. Writers may also determine that some of the ideas generated in the planning process need further elaboration, potentially necessitating the need for research to gather information to fill the gaps.

Once a writer has chosen his or her thesis and main argument, selected the most applicable details and evidence, and eliminated the "clutter," it is time to strategically organize the ideas. This is often accomplished with an outline.

Outlining

An **outline** is a system used to organize writing. When composing essays, outlining is important because it helps writers organize important information in a logical pattern using Roman numerals. Usually, outlines start out with the main ideas and then branch out into subgroups or subsidiary thoughts or subjects. Not only do outlines provide a visual tool for writers to reflect on how events, ideas, evidence, or other key parts of the argument relate to one another, but they can also lead writers to a stronger conclusion. The sample below demonstrates what a general outline looks like:

I. Introduction
     1. Background
     2. Thesis statement
II. Body
     1. Point A
          a. Supporting evidence
          b. Supporting evidence
     2. Point B
          a. Supporting evidence
          b. Supporting evidence
     3. Point C
          a. Supporting evidence
          b. Supporting evidence
III. Conclusion
     1. Restate main points of the paper.
     2. End with something memorable.

Drafting/Writing

Now it comes time to actually write the essay. In this stage, writers should follow the outline they developed in the brainstorming process and try to incorporate the useful sentences penned in the freewriting exercise. The main goal of this phase is to put all the thoughts together in cohesive sentences and paragraphs.

It is helpful for writers to remember that their work here does not have to be perfect. This process is often referred to as **drafting** because writers are just creating a rough draft of their work. Because of this, writers should avoid getting bogged down on the small details.

Revising

Revising offers an opportunity for writers to polish things up. Putting one's self in the reader's shoes and focusing on what the essay actually says helps writers identify problems—it's a movement from the mindset of writer to the mindset of editor. The goal is to have a clean, clear copy of the essay.

The main goal of the revision phase is to improve the essay's flow, cohesiveness, readability, and focus. For example, an essay will make a less persuasive argument if the various pieces of evidence are scattered and presented illogically or clouded with unnecessary thought. Therefore, writers should consider their essay's structure and organization, ensuring that there are smooth transitions between sentences and paragraphs. There should be a discernable introduction and conclusion as well, as these crucial components of an essay provide readers with a blueprint to follow.

Additionally, if the writer includes copious details that do little to enhance the argument, they may actually distract readers from focusing on the main ideas and detract from the strength of their work. The ultimate goal is to retain the purpose or focus of the essay and provide a reader-friendly experience.

Because of this, writers often need to delete parts of their essay to improve its flow and focus. Removing sentences, entire paragraphs, or large chunks of writing can be one of the toughest parts of the writing process because it is difficult to part with work one has done. However, ultimately, these types of cuts can significantly improve one's essay.

Lastly, writers should consider their voice and word choice. The voice should be consistent throughout and maintain a balance between an authoritative and warm style, to both inform and engage readers. One way to alter voice is through word choice. Writers should consider changing weak verbs to stronger ones and selecting more precise language in areas where wording is vague. In some cases, it is useful to modify sentence beginnings or to combine or split up sentences to provide a more varied sentence structure.

Editing

Rather than focusing on content (as is the aim in the revising stage), the editing phase is all about the mechanics of the essay: the syntax, word choice, and grammar. This can be considered the proofreading stage. Successful editing is what sets apart a messy essay from a polished document.

The following areas should be considered when proofreading:

- Sentence fragments
- Awkward sentence structure
- Run-on sentences
- Incorrect word choice
- Grammatical agreement errors
- Spelling errors
- Punctuation errors
- Capitalization errors

One of the most effective ways of identifying grammatical errors, awkward phrases, or unclear sentences is to read the essay out loud. Listening to one's own work can help move the writer from simply the author to the reader.

During the editing phase, it's also important to ensure the essay follows the correct formatting and citation rules as dictated by the assignment.

Practice Makes Prepared Writers

Like any other useful skill, writing only improves with practice. While writing may come more easily to some than others, it is still a skill to be honed and improved. Regardless of a person's natural abilities, there is always room for growth in writing. Practicing the basic skills of writing can aid in preparations for the exam.

One way to build vocabulary and enhance exposure to the written word is through reading. This can be through reading books, but reading of any materials such as newspapers, magazines, and even social media count towards practice with the written word. This also helps to enhance critical reading and thinking skills, through analysis of the ideas and concepts read. Think of each new reading experience as a chance to sharpen these skills.

Developing a Well-Organized Paragraph

A paragraph is a series of connected and related sentences addressing one topic. Writing good paragraphs benefits writers by helping them to stay on target while drafting and revising their work. It benefits readers by helping them to follow the writing more easily. Regardless of how brilliant their ideas may be, writers who do not present them in organized ways will fail to engage readers—and fail to accomplish their writing goals. A fundamental rule for paragraphing is to confine each paragraph to a single idea. When writers find themselves transitioning to a new idea, they should start a new paragraph. However, a paragraph can include several pieces of evidence supporting its single idea; and it can include several points if they are all related to the overall paragraph topic. When writers find each point becoming lengthy, they may choose instead to devote a separate paragraph to every point and elaborate upon each more fully.

An effective paragraph should have these elements:

- Unity: One major discussion point or focus should occupy the whole paragraph from beginning to end.

- Coherence: For readers to understand a paragraph, it must be coherent. Two components of coherence are logical and verbal bridges. In logical bridges, the writer may write consecutive sentences with parallel structure or carry an idea over across sentences. In verbal bridges, writers may repeat key words across sentences.

- A topic sentence: The paragraph should have a sentence that generally identifies the paragraph's thesis or main idea.

- Sufficient development: To develop a paragraph, writers can use the following techniques after stating their topic sentence:

  - Define terms
  - Cite data
  - Use illustrations, anecdotes, and examples
  - Evaluate causes and effects
  - Analyze the topic
  - Explain the topic using chronological order

A topic sentence identifies the main idea of the paragraph. Some are explicit, some implicit. The topic sentence can appear anywhere in the paragraph. However, many experts advise beginning writers to place each paragraph topic sentence at or near the beginning of its paragraph to ensure that their readers understand what the topic of each paragraph is. Even without having written an explicit topic sentence, the writer should still be able to summarize readily what subject matter each paragraph addresses. The writer must then fully develop the topic that is introduced or identified in the topic sentence. Depending on what the writer's purpose is, they may use different methods for developing each paragraph.

Two main steps in the process of organizing paragraphs and essays should both be completed after determining the writing's main point, while the writer is planning or outlining the work. The initial step is to give an order to the topics addressed in each paragraph. Writers must have logical reasons for putting one paragraph first, another second, etc. The second step is to sequence the sentences in each paragraph. As with the first step, writers must have logical reasons for the order of sentences. Sometimes the work's main point obviously indicates a specific order.

Topic Sentences
To be effective, a topic sentence should be concise so that readers get its point without losing the meaning among too many words. As an example, in *Only Yesterday: An Informal History of the 1920s* (1931), author Frederick Lewis Allen's topic sentence introduces his paragraph describing the 1929 stock

market crash: "The Bull Market was dead." This example illustrates the criteria of conciseness and brevity. It is also a strong sentence, expressed clearly and unambiguously. The topic sentence also introduces the paragraph, alerting the reader's attention to the main idea of the paragraph and the subject matter that follows the topic sentence.

Experts often recommend opening a paragraph with the topic sentences to enable the reader to realize the main point of the paragraph immediately. Application letters for jobs and university admissions also benefit from opening with topic sentences. However, positioning the topic sentence at the end of a paragraph is more logical when the paragraph identifies a number of specific details that accumulate evidence and then culminates with a generalization. While paragraphs with extremely obvious main ideas need no topic sentences, more often—and particularly for students learning to write—the topic sentence is the most important sentence in the paragraph. It not only communicates the main idea quickly to readers; it also helps writers produce and control information.

Frequently Misspelled Words

One source of spelling errors is not knowing whether to drop the final letter *e* from a word when its form is changed by adding an ending to indicate the past tense or progressive participle of a verb, converting an adjective to an adverb, a noun to an adjective, etc. Some words retain the final *e* when another syllable is added; others lose it. For example, *true* becomes *truly; argue* becomes *arguing; come* becomes *coming; write* becomes *writing;* and *judge* becomes *judging.* In these examples, the final *e* is dropped before adding the ending. But *severe* becomes *severely; complete* becomes *completely; sincere* becomes *sincerely; argue* becomes *argued;* and *care* becomes *careful.* In these instances, the final *e* is retained before adding the ending. Note that some words, like *argue* in these examples, drops the final *e* when the *–ing* ending is added to indicate the participial form; but the regular past tense ending of *–ed* makes it *argued*, in effect replacing the final *e* so that *arguing* is spelled without an *e* but *argued* is spelled with one.

Some English words contain the vowel combination of *ei,* while some contain the reverse combination of *ie.* Many people confuse these. Some examples include these:

> *ceiling, conceive, leisure, receive, weird, their, either, foreign, sovereign, neither, neighbors, seize, forfeit, counterfeit, height, weight, protein,* and *freight*

Words with *ie* include *piece, believe, chief, field, friend, grief, relief, mischief, siege, niece, priest, fierce, pierce, achieve, retrieve, hygiene, science,* and *diesel.* A rule that also functions as a mnemonic device is "I before E except after C, or when sounded like A as in 'neighbor' or 'weigh'." However, it is obvious from the list above that many exceptions exist.

Many people often misspell certain words by confusing whether they have the vowel *a, e,* or *i,* frequently in the middle syllable of three-syllable words or beginning the last syllables that sound the same in different words. For example, in the following correctly spelled words, the vowel in boldface is the one people typically get wrong by substituting one or either of the others for it:

> cem**e**tery, quant**i**ties, ben**e**fit, priv**i**lege, unpleas**a**nt, sep**a**rate, independ**e**nt, excell**e**nt, cat**e**gories, indispens**a**ble, and irrelev**a**nt

The words with final syllables that sound the same when spoken but are spelled differently include *unpleasant, independent, excellent,* and *irrelevant.* Another source of misspelling is whether or not to double consonants when adding suffixes. For example, we double the last consonant before *–ed* and *–ing* endings in *controlled, beginning, forgetting, admitted, occurred, referred,* and *hopping;* but we do not double the last consonant before the suffix in *shining, poured, sweating, loving, hating, smiling,* and *hoping.*

One way in which people misspell certain words frequently is by failing to include letters that are silent. Some letters are articulated when pronounced correctly but elided in some people's speech, which then transfers to their writing. Another source of misspelling is the converse: people add extraneous letters. For example, some people omit the silent *u* in *guarantee,* overlook the first *r* in *surprise,* leave out the *z* in *realize,* fail to double the *m* in *recommend,* leave out the middle *i* from *aspirin,* and exclude the *p* from *temperature.* The converse error, adding extra letters, is common in words like *until* by adding a second *l* at the end; or by inserting a superfluous syllabic *a* or *e* in the middle of *athletic,* reproducing a common mispronunciation.

Practice Essay

Thirty minutes is given to write your essay. Length is not as important as the quality of the writing, so focus on answering the question completely. Choose one of the three questions below to write your essay:

- Where is your favorite place to go and why?
- If you could meet anyone in the world, who would you meet and why?
- When was a time in your life when you felt excited? Describe the experience.

Practice Test

Verbal Reasoning

Synonyms

1. WARY
   a. Religious
   b. Adventurous
   c. Tired
   d. Cautious

2. PROXIMITY
   a. Estimate
   b. Delicate
   c. Closeness
   d. Splendor

3. STRIFE
   a. Plague
   b. Industrial
   c. Conflict
   d. Eliminate

4. QUALM
   a. Calm
   b. Uneasiness
   c. Assertion
   d. Pacify

5. VALOR
   a. Rare
   b. Coveted
   c. Leadership
   d. Bravery

6. ZEAL
   a. Craziness
   b. Resistance
   c. Fervor
   d. Opposition

7. REPROACH
   a. Locate
   b. Blame
   c. Concede
   d. Honor

8. GUILE
   a. Masculine
   b. Stubborn
   c. Naïve
   d. Deception

9. ASSENT
   a. Heighten
   b. Climb
   c. Assert
   d. Demand

10. DEARTH
    a. Grounded
    b. Scarcity
    c. Lethal
    d. Risky

11. CONSPICUOUS
    a. Scheme
    b. Obvious
    c. Secretive
    d. Paranoid

12. ONEROUS
    a. Responsible
    b. Generous
    c. Hateful
    d. Burdensome

13. BANAL
    a. Boring
    b. Novel
    c. Painful
    d. Complimentary

14. ASHAMED
    a. Tidy
    b. Unrealistic
    c. Remorseful
    d. Corrupt

15. CAPRICIOUS
    a. Skillful
    b. Agreeable
    c. Chaotic
    d. Fickle

16. PALTRY
    a. Appealing
    b. Worthy
    c. Trivial
    d. Fancy

17. SHIRK
    a. Counsel
    b. Evade
    c. Diminish
    d. Sharp

Sentence Completion

18. The painter was extremely _____; she had won three awards and sold thousands of paintings in her lifetime.
    a. talented
    b. mischievous
    c. boring
    d. lively

19. Olivia was feeling _____ from the car ride, so she closed her eyes and drifted off to sleep.
    a. ordinary
    b. harsh
    c. drowsy
    d. effortless

20. The place that Carlos arrived at was _____—he had never seen it before.
    a. anxious
    b. unfamiliar
    c. vertical
    d. cultured

21. We wanted to _____ the winners on stage before the reception began.
    a. dupe
    b. express
    c. hesitate
    d. announce

22. In English class we had to _____ which was the main character of the text in the novel we were reading.
    a. revolt
    b. dominate
    c. identify
    d. oppose

23. The _____ source means a source that was created at the same time that is being studied or analyzed. Some examples include a diary, recording, or artifact.
  a. accurate
  b. consistent
  c. dissatisfied
  d. primary

24. The comedian was _____. We laughed throughout the entire show.
  a. frequent
  b. hilarious
  c. harsh
  d. scarce

25. When the new dog from next door appeared in Harry's front yard, Harry _____ stuck his hand out to pet him.
  a. portably
  b. cautiously
  c. morally
  d. seldomly

26. Abigail never showed up at the correct time and was therefore not very _____.
  a. reliable
  b. frigid
  c. visual
  d. absurd

27. We couldn't tell the exact number of guests, but we knew it was _____ eighty.
  a. exactly
  b. sadly
  c. approximately
  d. occasionally

28. We loaded the _____ into the ship to be delivered in one week.
  a. riot
  b. morsel
  c. cargo
  d. voyage

29. While staying at the ski resort, we saw a great cliff of snow break off the mountain and discovered that it was a(n) _____.
  a. beverage
  b. jubilee
  c. catapult
  d. avalanche

30. Our vacation was very _____; we sat under palm trees on the ocean, watched kids playing with coconuts, and soaked under the warm, humid sun.
    a. tropical
    b. cold
    c. urban
    d. minor

31. At the _____, they served food and gave speeches.
    a. hospital
    b. prelude
    c. banquet
    d. hangar

32. The queen's _____ was prosperous for the country and lasted for fifty years.
    a. reign
    b. purchase
    c. harbor
    d. oasis

33. She was a(n) _____ because she believed in hope and favorable outcomes.
    a. investigator
    b. optimist
    c. journalist
    d. nihilist

34. The island was very _____; there were only five houses on it, and one had to get there by boat.
    a. formal
    b. colonial
    c. dormant
    d. remote

Quantitative Reasoning

1. The sum of 3 consecutive odd numbers is 183. What is the largest number?
    a. 57
    b. 69
    c. 63
    d. 59

$$X \quad X+2 \quad X+4$$
$$3X+6 = 183$$
$$3X = 177$$
$$X = \frac{177}{3} = 59, 61, 63$$

2. Carl makes $5 for every birdhouse he makes. Which equation shows the money, y, that he makes for a total number of birdhouses, b? $5 - 1$ $y: 5 * b$
    a. $Y = b - 5$
    b. $Y = 5 - b$
    c. $Y = 5b$
    d. $B = 5y$

3. The picture shows the shape of Ellie's yard, enclosed by a fence. What is the area of the yard inside the fence?

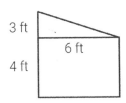

    a. 72 square feet
    b. 45 square feet
    c. 33 square feet
    d. 42 square feet

4. Which property is shown below?
$$x + y = y + x$$
    a. Commutative
    b. Distributive
    c. Associative
    d. Identity

5. If the points (1, 4), (5, 4), (1, 8), and (5, 8) are plotted on a coordinate plane, what shape do they form?
    a. Square
    b. Triangle
    c. Rectangle
    d. Trapezoid

6. Which story below fits the equation $3 \times 12 = 36$?
    a. Barry bought 36 balls for $3 each.
    b. Shelley had $12 to spend on 3 toys.
    c. Max scored 3 goals that were worth 12 points each.
    d. Lori bought 4 bowls of soup for $3 each.

7. There is an integer that is greater than negative two and less than four. It is two more than one less than zero. What is the integer?
    a. 0
    b. 1
    c. 2
    d. 3

8. The perimeter of the farmer's pasture depicted below is 32 centimeters. How long is the side?

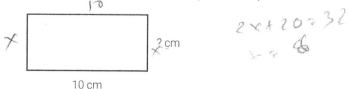

10 cm

    a. 4
    b. 5
    c. 6
    d. 7

9. Katie can run the same speed as Nolan. If Nolan ran 1 mile in 6 minutes, how long would it take Katie to run 2 miles?
    a. 9 minutes
    b. 15 minutes
    c. 6 minutes
    d. 12 minutes

10. Cain sells apples for \$5 more than Jackson. If Jackson sells apples for \$$d$, which equation shows the price Cain charges for his apples?
    a. $C = d + 5$
    b. $C = 5d$
    c. $C = 5 - d$
    d. $D = \dfrac{d}{5}$

11. What is the population of Indonesia and the United States combined?

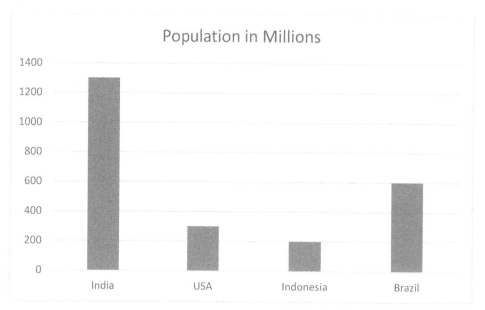

    a. 300 million
    b. 500 million
    c. 1400 million
    d. 200 million

12. If there are 6 tables needed for Shelley and Kyle's reception, how many chairs will be needed?

Number of Tables	Number of Chairs
1	4
2	7
3	9
4	11
5	13
6	14

  a. 12
  b. 24
  c. 14
  d. 8

13. Which property is shown below?

$$v(x + b) = vx + vb$$

  a. Commutative
  b. Distributive
  c. Associative
  d. Identity

14. Kori has 4.8 boxes of oranges that cost $3.10 per box. Approximately how much did she pay for her boxes of oranges?
  a. $12
  b. $15
  c. $16
  d. $20

15. Johnny has a bag that contains small game pieces of different shapes. He draws the following pieces out of the bag. What fraction of the pieces he drew are circles?

○ □ △ ○ ○ ○ ○ △ □

  a. $\frac{4}{7}$

  b. $\frac{3}{7}$

  c. $\frac{1}{2}$

  d. $\frac{3}{8}$

16. What shape is formed by the coordinates below?

$$(0, 2)(5, 2)(3, 7)$$

    a. Square
    b. Rectangle
    c. Triangle
    d. Parallelogram

17. There are 4 red marbles, 6 blue marbles, and 2 green marbles in a bag. What is the probability that Chase draws a blue marble from the bag?

    a. $\dfrac{1}{2}$

    b. $\dfrac{3}{4}$

    c. $\dfrac{6}{10}$

    d. $\dfrac{4}{10}$

18. What is the value of x in the flag pictured below?

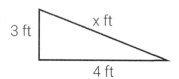

    a. 5 ft
    b. 12 ft
    c. 13 ft
    d. 7 ft

19. If the length of Fido's leash is 125 cm, how many meters long is it?
    a. 12.5 m
    b. 1.25 m
    c. 2.5 m
    d. 0.125 m

20. Choose the expression that gives the space covered by the following rectangular comic book.

    a. 5×10
    b. $5 + 10 + 5 + 10$
    c. $10 + 5$
    d. $2 \times 5 + 2 \times 10$

21. If the figure below is a rectangular box for chocolates, what is *e*?

    a. 45
    b. 90
    c. 180
    d. 60

22. According to the graph, how many students played ice hockey?

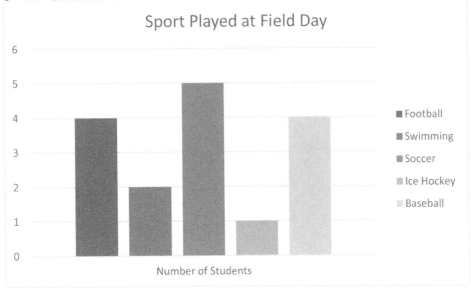

    a. 3 students
    b. 5 students
    c. 1 student
    d. 4 students

23. According to the graph, what was the distance traveled in 90 minutes?

a. 150 km
b. 100 km
c. 200 km
d. 90 km

24. What type of relationship is there between age and attention span as represented in the graph below?

Attention Span

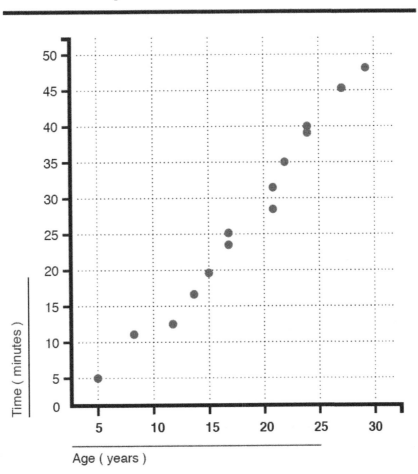

a. No correlation
b. Positive correlation
c. Negative correlation
d. Weak correlation

25. The following set represents the test scores from a university class: {35, 79, 80, 87| 87, 90, 92,|95, 95, 98,|99}. If the outlier is removed from this set, which of the following is TRUE?
    a. The mean and the median will decrease.
    b. The mean and the median will increase.
    c. The mean and the mode will increase.
    d. The mean and the mode will decrease.

26. Which of the statements below is a statistical question?
    a. What was your grade on the last test?
    b. What were the grades of the students in your class on the last test?
    c. What kind of car do you drive?
    d. What was Sam's time in the marathon?

27. Which shapes could NOT be used to compose a hexagon?

 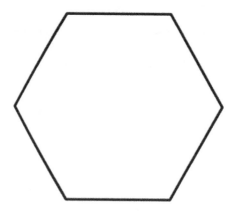

    a. Six triangles
    b. One rectangle and two triangles
    c. Two rectangles
    d. Two trapezoids

28. Two consecutive integers exist such that the sum of three times the first and two less than the second is equal to 411. What are those integers?
    a. 103 and 104
    b. 104 and 105
    c. 102 and 103
    d. 100 and 101

29. In a neighborhood, 15 out of 80 of the households have children under the age of 18. What percentage of the households have children?
    a. 0.1875%
    b. 18.75%
    c. 1.875%
    d. 15%

30. Gina took an algebra test last Friday. There were 35 questions, and she answered 60% of them correctly. How many correct answers did she have?
    a. 35
    b. 20
    c. 21
    d. 25

31. Paul took a written driving test, and he got 12 of the questions correct. If he answered 75% of the total questions correctly, how many problems were there in the test?
    a. 25
    b. 16
    c. 20
    d. 18

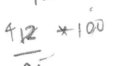 

114

32. If a car is purchased for $15,395 with a 7.25% sales tax, how much is the total price?
    a. $15,395.07
    b. $16,511.14
    c. $16,411.13
    d. $15,402

33. A car manufacturer usually makes 15,412 SUVs, 25,815 station wagons, 50,412 sedans, 8,123 trucks, and 18,312 hybrids a month. About how many cars are manufactured each month?
    a. 120,000
    b. 200,000
    c. 300,000
    d. 12,000

34. Each year, a family goes to the grocery store every week and spends $105. About how much does the family spend annually on groceries?
    a. $10,000
    b. $50,000
    c. $500
    d. $5,000

35. Bindee is having a barbeque on Sunday and needs 12 packets of ketchup for every 5 guests. If 60 guests are coming, how many packets of ketchup should she buy?
    a. 100
    b. 12
    c. 144
    d. 60

36. A grocery store sold 48 bags of apples in one day, and 9 of the bags contained Granny Smith apples. The rest contained Red Delicious apples. What is the ratio of bags of Granny Smith to bags of Red Delicious that were sold?
    a. 48:9
    b. 39:9
    c. 9:48
    d. 9:39

37. If Oscar's bank account totaled $4,000 in March and $4,900 in June, what was the rate of change in his bank account total over those three months?
    a. $900 a month
    b. $300 a month
    c. $4,900 a month
    d. $100 a month

38. Erin and Katie work at the same ice cream shop. Together, they always work less than 21 hours a week. In a week, if Katie worked two times as many hours as Erin, how many hours could Erin work?
    a. Less than 7 hours
    b. Less than or equal to 7 hours
    c. More than 7 hours
    d. Less than 8 hours

115

Reading Comprehension

Questions 1–5 are based on the following passage from The Story of Germ Life *by Herbert William Conn:*

When we study more carefully the effect upon the milk of the different species of bacteria found in the dairy, we find that there is a great variety of changes which they produce when they are allowed to grow in milk. The dairyman experiences many troubles with his milk. It sometimes curdles without becoming acid. Sometimes it becomes bitter, or acquires an unpleasant "tainted" taste, or, again, a "soapy" taste. Occasionally a dairyman finds his milk becoming slimy, instead of souring and curdling in the normal fashion. At such times, after a number of hours, the milk becomes so slimy that it can be drawn into long threads. Such an infection proves very troublesome, for many a time it persists in spite of all attempts made to remedy it. Again, in other cases the milk will turn blue, acquiring about the time it becomes sour a beautiful sky-blue colour. Or it may become red, or occasionally yellow. All of these troubles the dairyman owes to the presence in his milk of unusual species of bacteria which grow there abundantly.

1. What is the author's purpose in writing this passage?
   a. To show the readers that dairymen have difficult jobs
   b. To show the readers different ways their milk might go bad
   c. To show some of the different effects of milk on bacteria
   d. To show some of the different effects of bacteria on milk

2. What is the tone of this passage?
   a. Excitement
   b. Anger
   c. Neutral
   d. Sorrowful

3. Which of the following reactions does NOT occur in the above passage when bacteria infect the milk?
   a. It can have a soapy taste.
   b. The milk will turn black.
   c. It can become slimy.
   d. The milk will turn blue.

4. What is the meaning of "curdle" as depicted in the following sentence?
"Occasionally a dairyman finds his milk becoming slimy, instead of souring and <u>curdling</u> in the normal fashion."
   a. Lumpy
   b. Greasy
   c. Oily
   d. Slippery

5. Why, according to the passage, does an infection with slimy threads prove very troublesome?
   a. Because it is impossible to get rid of.
   b. Because it can make the milk-drinker sick.
   c. Because it turns the milk a blue color.
   d. Because it makes the milk taste bad.

Questions 6–10 are based on the passage from Many Marriages *by Sherwood Anderson:*

There was a man named Webster lived in a town of twenty-five thousand people in the state of Wisconsin. He had a wife named Mary and a daughter named Jane and he was himself a fairly prosperous manufacturer of washing machines. When the thing happened of which I am about to write he was thirty-seven or thirty-eight years old and his one child, the daughter, was seventeen. Of the details of his life up to the time a certain revolution happened within him it will be unnecessary to speak. He was however a rather quiet man inclined to have dreams which he tried to crush out of himself in order that he function as a washing machine manufacturer; and no doubt, at odd moments, when he was on a train going some place or perhaps on Sunday afternoons in the summer when he went alone to the deserted office of the factory and sat several hours looking out at a window and along a railroad track, he gave way to dreams.

6. What does the author mean by the following sentence?
   "Of the details of his life up to the time a certain revolution happened within him it will be unnecessary to speak."

   a. The details of his external life don't matter; only the details of his internal life matter.
   b. Whatever happened in his life before he had a certain internal change is irrelevant.
   c. He had a traumatic experience earlier in his life that rendered it impossible for him to speak.
   d. Before the revolution, he was a lighthearted man who always wished to speak to others no matter who they were.

7. From what Point Of View is this narrative told?
   a. First person limited
   b. First person omniscient
   c. Second person
   d. Third person

8. What did Webster do for a living?
   a. Washing machine manufacturer
   b. Train operator
   c. Leader of the revolution
   d. Stay-at-home husband

9. What does the word *deserted* mean?
   a. abandoned
   b. present
   c. lukewarm
   d. fundamental

10. What does the word *prosperous* mean?
    a. obedient
    b. desperate
    c. successful
    d. overdue

Questions 11–15 are based on the following passage. It is from Oregon, Washington, and Alaska. Sights and Scenes for the Tourist, *written by E.L. Lomax in 1890:*

Portland is a very beautiful city of 60,000 inhabitants, and situated on the Willamette river twelve miles from its junction with the Columbia. It is perhaps true of many of the growing cities of the West, that they do not offer the same social advantages as the older cities of the East. But this is principally the case as to what may be called boom cities, where the larger part of the population is of that floating class which follows in the line of temporary growth for the purposes of speculation, and in no sense applies to those centers of trade whose prosperity is based on the solid foundation of legitimate business. As the metropolis of a vast section of country, having broad agricultural valleys filled with improved farms, surrounded by mountains rich in mineral wealth, and boundless forests of as fine timber as the world produces, the cause of Portland's growth and prosperity is the trade which it has as the center of collection and distribution of this great wealth of natural resources, and it has attracted, not the boomer and speculator, who find their profits in the wild excitement of the boom, but the merchant, manufacturer, and investor, who seek the surer if slower channels of legitimate business and investment. These have come from the East, most of them within the last few years. They came as seeking a better and wider field to engage in the same occupations they had followed in their Eastern homes, and bringing with them all the love of polite life which they had acquired there, have established here a new society, equaling in all respects that which they left behind. Here are as fine churches, as complete a system of schools, as fine residences, as great a love of music and art, as can be found at any city of the East of equal size.

11. What is a characteristic of a "boom city," as indicated by the passage?
    a. A city that is built on solid business foundation of mineral wealth and farming
    b. An area of land on the west coast that quickly becomes populated by residents from the east coast
    c. A city that, due to the hot weather and dry climate, catches fire frequently, resulting in a devastating population drop
    d. A city whose population is made up of people who seek quick fortunes rather than building a solid business foundation

12. The author would classify Portland as which of the following?
    a. A boom city
    b. A city on the east coast
    c. An industrial city
    d. A city of legitimate business

13. What type of passage is this?
    a. A business proposition
    b. A travel guide
    c. A journal entry
    d. A scholarly article

14. What does the word *metropolis* mean in the middle of the passage?
    a. Farm
    b. Country
    c. City
    d. Valley

118

15. What does the word *legitimate* mean?
    a. employ
    b. donate
    c. obstacle
    (d) authentic

Questions 16–20 are based on the following passage from the biography Queen Victoria *by E. Gordon Browne, M.A.:*

> The old castle soon proved to be too small for the family, and in September 1853 the foundation-stone of a new house was laid. After the ceremony, the workmen were entertained at dinner, which was followed by Highland games and dancing in the ballroom.
>
> Two years later, they entered the new castle, which the Queen described as "charming; the rooms delightful; the furniture, papers, everything perfection."
>
> The Prince was untiring in planning improvements, and in 1856 the Queen wrote: "Every year my heart becomes more fixed in this dear Paradise, and so much more so now, that *all* has become my dearest Albert's *own* creation, own work, own building, own laying out as at Osborne; and his great taste, and the impress of his dear hand, have been stamped everywhere. He was very busy today, settling and arranging many things for next year."

16. This excerpt is considered which of the following?
    a. Primary source
    b. Secondary source
    c. Tertiary source
    (d) None of these

17. How many years did it take for the new castle to be built?
    a. One year
    (b) Two years
    c. Three years
    d. Four years

18. What does the word *impress* mean in the third paragraph?
    a. To affect strongly in feeling
    b. To urge something to be done
    (c) To impose a certain quality upon
    d. To press a thing onto something else

19. When was the foundation of the new house laid?
    a. 1849
    b. 1850
    (c) 1853
    d. 1856

20. What is the tone of this passage?
    a. Indifferent
    b. Excited
    c. Anxious
    d. Depressing

Questions 21–25 are based on the following passage from A Christmas Carol *by Charles Dickens:*

> Meanwhile the fog and darkness thickened so, that people ran about with flaring links, proffering their services to go before horses in carriages, and conduct them on their way. The ancient tower of a church, whose gruff old bell was always peeping slyly down at Scrooge out of a Gothic window in the wall, became invisible, and struck the hours and quarters in the clouds, with tremulous vibrations afterwards as if its teeth were chattering in its frozen head up there. The cold became intense. In the main street, at the corner of the court, some labourers were repairing the gas-pipes, and had lighted a great fire in a brazier, round which a party of ragged men and boys were gathered: warming their hands and winking their eyes before the blaze in rapture. The water-plug being left in solitude, its overflowings sullenly congealed, and turned to misanthropic ice. The brightness of the shops where holly sprigs and berries crackled in the lamp heat of the windows, made pale faces ruddy as they passed. Poulterers' and grocers' trades became a splendid joke; a glorious pageant, with which it was next to impossible to believe that such dull principles as bargain and sale had anything to do. The Lord Mayor, in the stronghold of the mighty Mansion House, gave orders to his fifty cooks and butlers to keep Christmas as a Lord Mayor's household should; and even the little tailor, whom he had fined five shillings on the previous Monday for being drunk and bloodthirsty in the streets, stirred up to-morrow's pudding in his garret, while his lean wife and the baby sallied out to buy the beef.

21. In the context in which it appears, *congealed* most nearly means which of the following?
    a. Burst
    b. Loosened
    c. Shrank
    d. Thickened

22. Which of the following can NOT be inferred from the passage?
    a. The season of this narrative is in the wintertime.
    b. The majority of the narrative is located in a bustling city street.
    c. This passage takes place during the nighttime.
    d. The Lord Mayor is a wealthy person within the narrative.

23. According to the passage, which of the following is true about the poulterers and grocers?
    a. They were so poor in the quality of their products that customers saw them as a joke.
    b. They put on a pageant in the streets every year for Christmas to entice their customers.
    c. They did not believe in Christmas, so they refused to participate in the town parade.
    d. They set their shops up to be entertaining public spectacles rather than a dull trade exchange.

24. What is the meaning of the word *proffering* in this passage?
    a. Giving away
    b. Offering
    c. Bolstering
    d. Teaching

25. What does the author mean by the following sentence?
The brightness of the shops where holly sprigs and berries crackled in the lamp heat of the windows, made pale faces ruddy as they passed.

a. When people walked past the shops, their faces turned red because of the lamps in the windows that were also lighting up holly sprigs and berries.
b. Compared with the holly sprigs and berries and their crackling lamplight, everyone's face looked old when they walked by the shops.
c. When people walked past the shops, their faces looked cold and blue compared with the warm light of the shops, which were making the holly sprigs and berries glow.
d. While shop owners were cooking their holly sprigs and berries in the warm glow of the fire, people's faces lit up with excitement as they passed.

Mathematics Achievement

1. What is $\frac{12}{60}$ converted to a percentage?
   a. 0.20
   b. 20%
   c. 25%
   d. 12%

$$0.20$$
$$60\overline{|12.00}$$
$$0$$
$$12\ 0$$

$$0.20 \times 100$$
$$20\ \%$$

2. Which of the following is the correct decimal form of the fraction $\frac{14}{33}$ rounded to the nearest hundredth place?
   a. 0.420
   b. 0.42
   c. 0.424
   d. 0.140

$$.4214$$
$$.420$$

$$0.4214$$
$$33\overline{|14.000}$$
$$0$$
$$140$$
$$132$$
$$80$$
$$66$$
$$140$$

3. Which of the following represents the correct sum of $\frac{14}{15}$ and $\frac{2}{5}$?
   a. $\frac{20}{15}$

   b. $\frac{4}{3}$

   c. $\frac{16}{20}$

   d. $\frac{4}{5}$

$$\frac{14}{15} + \frac{6}{15} = \frac{20}{15}$$

✓ 4. What is the product of $\frac{5}{14}$ and $\frac{7}{20}$? $= \frac{1}{8}$

(a.) $\frac{1}{8}$

b. $\frac{35}{280}$

c. $\frac{12}{34}$

d. $\frac{1}{2}$

✓ 5. What is the result of dividing 24 by $\frac{8}{5}$?

a. $\frac{5}{3}$

b. $\frac{3}{5}$

c. $\frac{120}{8}$

(d.) 15

✓ 6. Subtract $\frac{5}{14}$ from $\frac{5}{24}$. Which of the following is the correct result?

a. $\frac{25}{168}$

b. 0

(c.) $-\frac{25}{168}$

d. $\frac{1}{10}$

✓ 7. Eva Jane is practicing for an upcoming 5K run. She has recorded the following times (in minutes):
25, 18, 23, 28, 30, 22.5, 23, 33, 20
Use the above information to answer the next three questions to the closest minute. What is Eva Jane's mean time?

a. 26 minutes

b. 19 minutes

(c.) 25 minutes

d. 23 minutes

✓ 8. What is the mode of Eva Jane's time?

a. 16 minutes

b. 20 minutes

(c.) 23 minutes

d. 33 minutes

✓ 9. What is Eva Jane's median time?

(a.) 23 minutes

b. 17 minutes

c. 28 minutes

d. 19 minutes

18, 20, 23, 23, 25, 26, 28, 30, 34

122

10. What is the solution to the equation $10 - 5x + 2 = 7x + 12 - 12x$? $10 + 2 - 12 = 5x = 7x + 12x$
   a. $x = 12$
   b. No solution
   c. $x = 0$
   d. All real numbers

   $12 - 12 = 5x = 7x + 12x$

   $0 = 5x = 7x + 12x =$

11. Which of the following is the result when solving the equation $4(x + 5) + 6 = 2(2x + 3)$?
   a. Any real number is a solution.
   b. There is no solution.
   c. $x = 6$ is the solution.
   d. $x = 26$ is the solution.

   $4 \times 11 = 44 + 6 = 50$ $31 \times 4 = 124 + 6 = 130$

   $2 \times (2 \times 6 + 3) = 30$ $2(2 \times 26 + 8) = 60 =$

   120

12. How many cases of cola can Lexi purchase if each case is $3.50 and she has $40?
   a. 10
   b. 12
   c. 11.4
   d. 11

   $\begin{array}{r} \times\ 3.50 \\ 10 \\ \hline 0\ 00 \\ 350 \\ \hline 3\ 5.00 \\ +\ 3.50 \\ \hline 38.50 \end{array}$

13. What is the area of the non-shaded region?

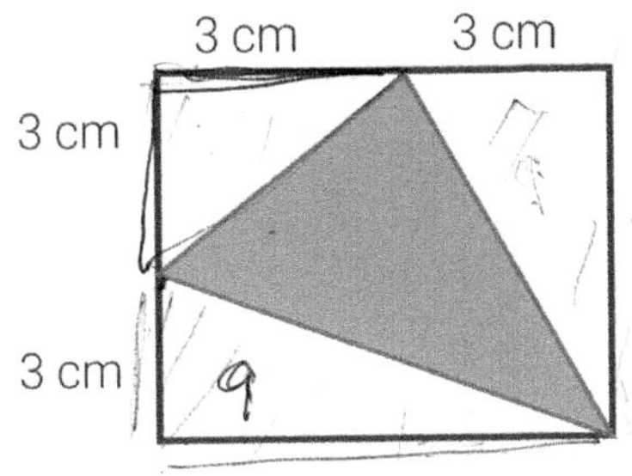

   3 cm   3 cm

   3 cm

   3 cm   9

   a. 45 cm
   b. 24 cm
   c. 30 cm
   d. 22.5 cm

   6

   $\begin{array}{r} 18 \\ -\ 2 \\ +\ 9 \\ \hline 45 \end{array}$

   36 36 18

14. Find $2 \times 3 \times 25 \times 4 \times 10$.
   a. 600
   b. 6,000
   c. 7,500
   d. 400

   $\dfrac{1 + 6 \times 3 = 5}{2}$

   $\dfrac{1 \times 3 \times 3}{2} = 4.5$ 22.5

15. What fraction of the shapes consists of triangles?

   $\square\ \bigcirc\ \triangle\ \square\ \square\ \triangle\ \triangle\ \triangle$

   a. $\dfrac{2}{3}$
   b. $\dfrac{4}{7}$
   c. $\dfrac{1}{2}$
   d. $\dfrac{2}{7}$

16. Estimate 4.9 × 3.4.
   a. 20
   b. 15
   c. 18
   d. 17.9

17. What fraction of the shape is shaded?

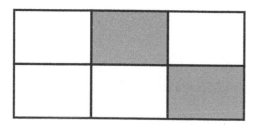

   a. $\frac{2}{3}$

   b. $\frac{1}{3}$

   c. $\frac{4}{5}$

   d. $\frac{1}{6}$

18. If $3 + x = 12$ and $4 + y = 10$, then what is $x + y$?
   a. 9
   b. 6
   c. 12
   d. 15

$12 - 3 \qquad 10 - 4$

$9 + 6 = 15$

19. What is the value of x?

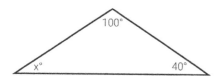

   a. 20 degrees
   b. 60 degrees
   c. 40 degrees
   d. 30 degrees

20. Which of the following is a correct mathematical statement?
   a. $\frac{1}{3} < -\frac{4}{3}$

   b. $-\frac{1}{3} > \frac{4}{3}$

   c. $\frac{1}{3} > -\frac{4}{3}$

   d. $-\frac{1}{3} \geq \frac{4}{3}$

124

21. Which of the following is incorrect?

a. $-\frac{1}{5} < \frac{4}{5}$

b. $\frac{4}{5} > -\frac{1}{5}$

c. $-\frac{1}{5} > \frac{4}{5}$

d. $\frac{1}{5} > -\frac{4}{5}$

22. Which type of graph best represents a continuous change over a period of time?

a. Bar graph

b. Line graph

c. Pie graph

d. Histogram

23. Which equation correctly shows how to find the volume the following cylindrical soup can?

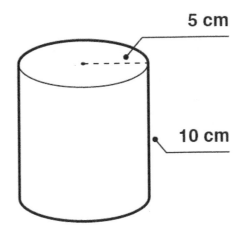

5 cm

10 cm

a. $V = 2\pi \times 5 \times 10$

b. $V = \pi \times 5 \times 10$

c. $V = \pi \times 5^2 \times 10$

d. $V = 2\pi \times 5^2 \times 10$

24. From the chart below, which two are preferred by more men than women?

Preferred Movie Genres

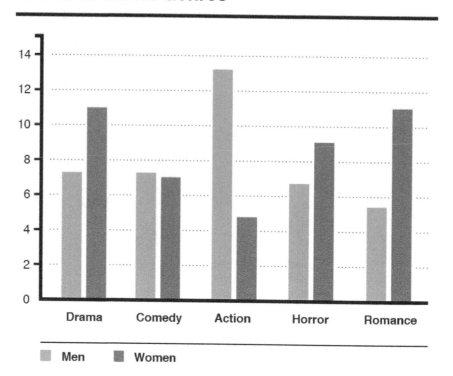

a. Comedy and Action
b. Drama and Comedy
c. Action and Horror
d. Action and Romance

25. What is the solution to the equation $3(x + 2) = 14x - 5$?
a. $x = 1$
b. No solution
c. $x = 0$
d. All real numbers

26. Using the graph below, what is the mean number of visitors for the first 4 hours?

Museum Visitors

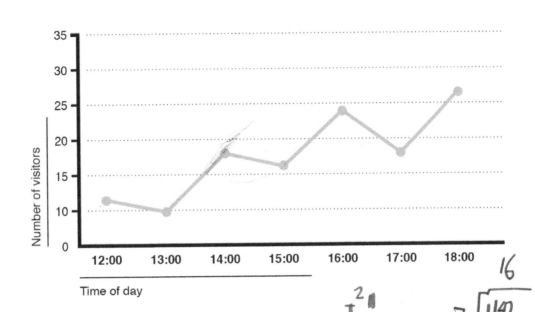

Time of day

a. 12
b. 13
c. 14
d. 15

$$7\overline{)112}$$ = 16

2
$+2$
10
$+16$
15
24
13
21
112

16
$\times 16$
7
112

27. What is the mode for the grades shown in the chart below?

Science Grades	
Jerry	65
Bill	95
Anna	80
Beth	95
Sara	85
Ben	72
Jordan	98

a. 65
b. 33
c. 95
d. 90

127

X 28. What is the area of the shaded region?

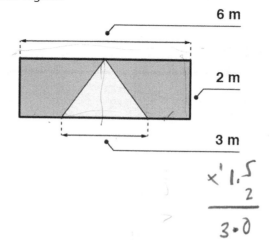

a. 9 m²
b. 12 m²
c. 6 m²
d. 8 m²

29. What is the volume of the cylinder below?

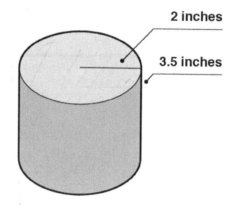

a. 18.84 in³
b. 45.00 in³
c. 70.43 in³
d. 43.96 in³

30. How many kiloliters are in 6 liters?
a. 6,000
b. 600
c. 0.006
d. 0.0006

Practice Essay

Thirty minutes is given to write your essay. Length is not as important as the quality of the writing, so focus on answering the question completely. Choose one of the three questions below to write your essay:

- Where is your favorite holiday and why?
- Who inspires you to be your best and why?
- When was a time in your life when you faced an obstacle? Describe the experience.

Answer Explanations
Verbal Reasoning

1. D: Someone who is *wary* is overly cautious or apprehensive. This word is often used in the context of being watchful or on guard about a potential danger. For example, darkening clouds and white caps on the waves may make a seaman wary against setting sail.

2. C: *Proximity* is defined as closeness, or the state or quality of being near in place, time, or relation.

3. C: *Strife* is a noun that is defined as bitter or vigorous discord, conflict, or dissension. It can mean a fight or struggle, or other act of contention. For example, antagonistic political interest groups vying for local support may be at strife.

4. B: *Qualm* is a noun that means a feeling of apprehension or uneasiness, often brought on suddenly. A girl who is just learning to ride a bike may have qualms about getting back on the saddle after taking a bad fall. It may also refer to an uneasy feeling related to one's conscience as it pertains to his or her actions. For example, a man with poor morals may have no qualms about lying on his tax return.

5. D: *Valor* is bravery or courage when facing a formidable danger. It often relates to strength of mind or spirit during battle or acting heroically in such situations.

6. C: *Zeal* is eagerness, fervor, or ardent desire in the pursuit of something. For example, a competitive collegiate baseball player's zeal to succeed in his sport may compromise his academic performance.

7. B: To *reproach* means to blame, find fault, or severely criticize. It can also be defined as expressing significant disapproval. It is often used in the phrase "beyond reproach" as in, "her violin performance was beyond reproach." In this context, it means her playing was so good that it evaded any possibility of criticism.

8. D: *Guile* can be defined as the quality of being cunning or crafty and skilled in deception. Someone may use guile to trick or deceive someone, like to get money from them, or otherwise dupe them.

9. A: As a verb, *assent* means to express agreement, give consent, or acquiesce. A job candidate might assent to an interviewer's request to perform a background check. As a noun, it means an agreement, acceptance, or acquiescence.

10. B: *Dearth* means a lack of something or a scarcity or shortage. For example, a local library might have a dearth of information pertaining to an esoteric topic.

11. B: *Conspicuous* means to be visually or mentally obvious. Something conspicuous stands out, is clearly visible, or may attract attention.

12. D: *Onerous* most closely means burdensome or troublesome. It usually is used to describe a task or obligation that may impose a hardship or burden, often which may be perceived to outweigh its benefits.

13: A: Something *banal* lacks originality, and may be boring and trite. For example, a banal compliment is likely to be a common platitude. Like something that is inane, a banal compliment might be meaningless and lack a convincing quality or significance.

14. C: *Ashamed* is closest in meaning to remorseful. Someone who is ashamed of something is sorry or remorseful.

15. D: Of the provided choices, *fickle* is closest in meaning to *capricious*. A person who is capricious tends to display erratic or unpredictable behavior, which is similar to fickle, which is also likely to change spontaneously or behave erratically.

16. C: Although *paltry* often is used as an adjective to describe a very small or meager amount (of money, in particular), it can also mean something trivial or insignificant.

17. B: The word *shirk* means to evade, and is often used in the context of shirking a responsibility, duty, or work.

18. A: The painter was extremely *talented*. For a painter to win awards and sell their paintings, they have to be talented, or skilled at what they do, so people will buy their work.

19. C: Olivia was feeling *drowsy* from the car ride. If we look at the context clues in the surrounding sentence, we can tell that someone closing their eyes and drifting off to sleep indicates that they are feeling drowsy, or sleepy.

20. B: The place that Carlos arrived at was *unfamiliar*. To go to a place that one has never seen before means that the place would be unfamiliar, different, or strange.

21. D: We wanted to *announce* the winners. *Announce* is the best fit for the sentence because of the surrounding context. We see that there is a reception and that before the reception there is to be something done on stage. *Announce* is an action that can be done on stage, especially when it comes to someone winning something.

22. C: In English class we had to *identify* which was the main character of the text. Identify means to recognize or label something.

23. D: A *primary* source is one that was created at the same time that is being studied or analyzed. Primary means first or original.

24. B: The comedian was *hilarious*. The surrounding context shows that the speaker laughed at the comedian's show, which implies that the comedian was *hilarious*, or funny.

25. B: Harry *cautiously* stuck his hand out to pet him. To be cautious of something means to be hesitant or careful. Harry doesn't know if the new dog is friendly or mean, so he slowly, carefully, sticks out his hand.

26. A: Abigail was not very *reliable*. Reliable means trustworthy or dependable, so if Abigail does not show up on time, then she is not very reliable.

27. C: We knew it was *approximately* eighty. The word approximately means almost exact or almost accurate. Since they couldn't tell the *exact* number of guests, they guessed at the approximate number of guests.

28. C: We loaded the *cargo* into the ship. Cargo is baggage or shipment that a ship carries for delivery.

29. D: We discovered that it was an *avalanche*. An avalanche is a sudden rush of something, especially snow coming from a mountain. A ski resort surrounded by mountains is very likely to witness a snow avalanche.

30. A: The word *tropical* signifies characteristics of a hot, humid, temperate climate usually in the tropics. Tropical is the best word here given the context clues of weather, animals, and activities related to a tropical atmosphere.

31. C: The best word is *banquet*. Looking at the context clues, we have to choose a noun that involves a place where people serve food and give speeches, such as a banquet.

32. A: The queen's *reign* was prosperous for the country and lasted for fifty years. The word reign means the period of time in which a king or queen occupies a throne.

33. B: She was an *optimist* because she believed in hope and favorable outcomes. An optimist believes that everything will turn out in good favor.

34. D: The island was very *remote*. Remote means that something is secluded or far away from civilization.

Quantitative Reasoning

1. C: The three consecutive integers are numbers that lie one after another and they must all be odd. Because there are three numbers, a good start is to divide 183 by 3 which is 61. The odd number before this is 59 and the odd number after if it is 63. Adding 59, 61, and 63 results in a sum of 183, so the largest of these numbers is 63.

2. C: Since Carl makes $5 for each birdhouse and the number of birdhouses is b, then the money made can be found by multiplying 5 by b, or $y = 5b$.

3. C: The area of the yard inside the fence is found by separating it into a triangle and a rectangle. The triangle's area is found by the formula $A = \frac{1}{2} \times 3 \times 6 = 9$ square feet. The area of the rectangle is found by multiplying the side length 6 by the side length 4 to yield an area of 24 square feet. Adding 24 to 9 gives a total area of 33 square feet for the yard.

4. A: The commutative property is shown. The only thing that changes on each side of the equal sign is the order of the variables.

5. A: When these values are plotted on a coordinate plane, they form a square because the distance between points around right angles is the same.

6. C: Max scored 3 goals that were worth 12 points apiece, which gives him a total of 36 points.

7. B: One less than zero is negative one, and then two more than negative one is positive one. This problem can be worked out by working backwards.

8. C: The perimeter of a rectangle is found by adding up all the sides. Since one side of 10 cm is given, then the opposite side is also 10 cm. Subtracting the two sides of 10 from the perimeter of 32 cm leaves 12 cm for the other two sides. Taking half of 12 cm gives a side length of 6 cm.

9. D: Since Katie runs at the same speed at Nolan, then the time to run 1 mile can be doubled to find the time to run 2 miles. Doubling a time of 6 minutes gives a time of 12 minutes.

10. A: The equation $c = d + 5$ gives an equation for the cost of Cain's apples because he charges $5 more than Jackson.

11. B: The population of the United States is 300 million according to the vertical axis on the graph. The population of Indonesia is 200 million according to the vertical axis on graph. Adding these two populations yields a total population of 500 million.

12. C: They need 6 tables, and the graph shows that 14 chairs will be needed. These two values line up on the graph.

13. B: The distributive property obtains when the variable on the outside of the parenthesis is distributed to each variable inside the parenthesis.

14. B: Kori has approximately 5 boxes of oranges and they cost approximately $3 each, so the approximate value of $15 is found with the equation $3 \times 5 = 15$.

15. C: There are 4 circles out of a total of 8 shapes in the group. The fraction $\frac{4}{8}$ can be simplified to $\frac{1}{2}$.

16. C: These points form a triangle on the graph because there are three points, which when connected form a shape with three sides.

17. A: The probability of drawing a blue marble is 6 out of 12 since there are 6 blue marbles in a bag of 12 total marbles. The number 6 is half of 12, so the probability is $\frac{1}{2}$.

18. A: The measures of 3, 4, and 5 for triangle side lengths are common for right triangles. These values can be confirmed by plugging the numbers in to the Pythagorean theorem, $a^2 + b^2 = c^2$.

19. B: There are 100 centimeters in a meter, so the conversion from centimeters to meters means moving the decimal two places to the left. For a leash length of 125 cm, the length in meters is 1.25.

20. A: The area of a rectangle is found by multiplying the length times the width. The length of 10 inches multiplied by the width of 5 inches gives an area of 50 square inches, so the correct expression is 5×10.

21. B: Since this shape is a rectangle, the angle measurements inside it are all equal to 90 degrees.

22. C: According to the graph, there is only one student who plays ice hockey.

23. A: The distance that corresponds to 90 minutes on the graph is 150 km. First find the 90 minutes on the x-axis, then follow with your finger up to the corresponding y-value of 150 km.

24. B: The relationship between age and time for attention span is a positive correlation because the general trend for the data is up and to the right. As the age increases, so does attention span.

25. B: The outlier is 35. When a small outlier is removed from a data set, the mean and the median increase. The first step in this process is to identify the outlier, which is the number that lies away from the given set. Once the outlier is identified, the mean and median can be recalculated. The mean will be affected because it averages all of the numbers. The median will be affected because it finds the middle number, which is subject to change because a number is lost. The mode will most likely not change because it is the number that occurs the most, which will not be the outlier if there is only one outlier.

26. B: This is a statistical question because to determine this answer one would need to collect data from each person in the class and it is expected the answers would vary. The other answers do not require data to be collected from multiple sources, therefore the answers will not vary.

27. C: A hexagon can be formed by any combination of the given shapes except for two rectangles. There are no two rectangles that can make up a hexagon.

28. A: First, the variables have to be defined. Let x be the first integer; therefore, $x + 1$ is the second integer. This is a two-step problem. The sum of three times the first and two less than the second is translated into the following expression:

$$3x + (x + 1 - 2)$$

This expression is set equal to 411 to obtain:

$$3x + (x + 1 - 2) = 411$$

The left-hand side is simplified to obtain:

$$4x - 1 = 411$$

The addition and multiplication properties are used to solve for x. First, add 1 to both sides and then divide both sides by 4 to obtain $x = 103$. The next consecutive integer is 104.

29. B: First, the information is translated into the ratio $\frac{15}{80}$. To find the percentage, translate this fraction into a decimal by dividing 15 by 80. The corresponding decimal is 0.1875. Move the decimal point two units to the right to obtain the percentage 18.75%.

30. C: Gina answered 60% of 35 questions correctly; 60% can be expressed as the decimal 0.60. Therefore, she answered $0.60 \times 35 = 21$ questions correctly.

31. B: The unknown quantity is the number of total questions on the test. Let x be equal to this unknown quantity. Therefore, $0.75x = 12$. Divide both sides by 0.75 to obtain $x = 16$.

32. B: If sales tax is 7.25%, the price of the car must be multiplied times 1.0725 to account for the additional sales tax. Therefore:

$$15,395 \times 1.0725 = 16,511.1375$$

This amount is rounded to the nearest cent, which is $16,511.14.

33. A: Rounding can be used to find the best approximation. All of the values can be rounded to the nearest thousand. 15,412 SUVs can be rounded to 15,000. 25,815 station wagons can be rounded to 26,000. 50,412 sedans can be rounded to 50,000. 8,123 trucks can be rounded to 8,000. Finally, 18,312 hybrids can be rounded to 18,000. The sum of the rounded values is 117,000, which is closest to 120,000.

34. D: There are 52 weeks in a year, and if the family spends $105 each week, that amount is close to $100. A good approximation is $100 a week for 50 weeks, which is found through the product:

$$50 \times 100 = \$5,000$$

35. C: This problem involves ratios and percentages. If 12 packets are needed for every 5 people, this statement is equivalent to the ratio $\frac{12}{5}$. The unknown amount x is the number of ketchup packets needed for 60 people. The proportion $\frac{12}{5} = \frac{x}{60}$ must be solved. Cross-multiply to obtain:

$$12 \times 60 = 5x$$

Therefore, $720 = 5x$. Divide each side by 5 to obtain $x = 144$.

36. D: There were 48 total bags of apples sold. If 9 bags were Granny Smith and the rest were Red Delicious, then $48 - 9 = 39$ bags were Red Delicious. Therefore, the ratio of Granny Smith to Red Delicious is 9:39.

37. B: The average rate of change is found by calculating the difference in dollars over the elapsed time. Therefore, the rate of change is equal to ($\$4,900 - \$4,000$) ÷ 3 months, which is equal to $\$900 \div 3$ or $\$300$ per month.

38. A: Let x be the unknown, the number of hours Erin can work. We know Katie works $2x$, and the sum of all hours is less than 21. Therefore, $x + 2x < 21$, which simplifies into $3x < 21$. Solving this results in the inequality $x < 7$ after dividing both sides by 3. Therefore, Erin can work less than 7 hours.

Reading Comprehension

1. D: The passage explains the different ways bacteria can affect milk by detailing the color and taste of different infections. Choices *A* and *B* might be true, but they are not the main purpose of the passage as detailed in the first sentence. Choice *C* is incorrect, since milk does not have an effect on bacteria in this particular passage.

2. C: The tone of this passage is neutral, since it is written in an academic/informative voice. It is important to look at the author's word choice to determine the tone of a passage. We have no indication that the author is excited, angry, or sorrowful at the effects of bacteria on milk, so Choices *A, B,* and *D* are incorrect.

3. B: The milk will turn black is not a reaction mentioned in the passage. The passage does state, however, that the milk may get "soapy," that it can become "slimy," and that it may turn out to be a "beautiful sky-blue colour," making Choices *A, C,* and *D* incorrect.

4. A: In the sentence, we know that the word "curdle" means the opposite of "slimy." The words greasy, oily, and slippery are all very similar to the word slimy, making Choices *B, C,* and *D* incorrect. "Lumpy" means clotted, chunky, or thickened.

5. A: It is troublesome because it is impossible to get rid of. The passage mentions milk turning blue or tasting bad, and milk could possibly even make a milk-drinker sick if it has slimy threads. However, we know for sure that the slimy threads prove troublesome because they can become impossible to get rid of from this sentence: "Such an infection proves very troublesome, for many a time it persists in spite of all attempts made to remedy it."

6. B: The sentence is best taken to mean that whatever happened in his life before he had a certain internal change is irrelevant. Choices *A, C,* and *D* use some of the same language as the original passage, like "revolution," "speak," and "details," but they do not capture the meaning of the statement. The statement is saying the details of his previous life are not going to be talked about—that he had some kind of epiphany, and moving forward in his life is what the narrator cares about.

7. B: It is told in first-person omniscient. This is the best guess with the information we have. In the world of the passage, the narrator is first-person, because we see them use the "I," but they also know the actions and thoughts of the protagonist, a character named "Webster." First-person limited tells their own story, making Choice *A* incorrect. Choice *C* is incorrect; second person uses "you" to tell the story. Third person uses "them," "they," etc., and would not fall into use of the "I" in the narrative, making Choice *D* incorrect.

8. A: Webster is a washing machine manufacturer. This question assesses reading comprehension. We see in the second sentence that Webster "was a fairly prosperous manufacturer of washing machines," making Choice A the correct answer.

9. A: The word *deserted* means abandoned. The clue in the passage says that Webster would go *alone* to a *deserted* office, which means the office is empty or abandoned.

10. C: The word *prosperous* means successful or thriving. In the text, Webster is a man who is a *successful* manufacturer of washing machines.

11. D: A "boom city" is a city whose population is made up of people who seek quick fortunes rather than building a solid business foundation. Choice *A* is a characteristic of Portland, but not that of a boom city. Choice *B* is close—a boom city is one that becomes quickly populated, but it is not necessarily always populated by residents from the east coast. Choice *C* is incorrect because a boom city is not one that catches fire frequently, but one made up of people who are looking to make quick fortunes from the resources provided on the land.

12. D: The author would classify Portland as a city of legitimate business. We can see the proof in this sentence: "the cause of Portland's growth and prosperity is the trade which it has as the center of collection and distribution of this great wealth of natural resources, and it has attracted, not the boomer and speculator . . . but the merchant, manufacturer, and investor, who seek the surer if slower channels of legitimate business and investment." Choices *A, B*, and *C* are not mentioned in the passage and are incorrect.

13. B: This passage is part of a travel guide. Our first hint is in the title: *Oregon, Washington, and Alaska. Sights and Scenes for the Tourist*. Although the passage talks about business, there is no proposition included, which makes Choice *A* incorrect. Choice *C* is incorrect because the style of the writing is more informative and formal rather than personal and informal. Choice *D* is incorrect; this could possibly be a scholarly article, but the best choice is that it is a travel guide, due to the title and the details of what the city has to offer at the very end.

14. C: *Metropolis* means city. Portland is described as having agricultural valleys, but it is not solely a "farm" or "valley," making Choices *A* and *D* incorrect. We know from the description of Portland that it is more representative of a city than a countryside or country, making Choice *B* incorrect.

15. D: The word *legitimate* means authentic. In the text, those who come to invest or manufacture using Portland's resources seek *authentic* businesses rather than fast-paced temporary work.

16. B: This excerpt is considered a secondary source because it actively interprets primary sources. We see direct quotes from the queen, which would be considered a primary source. But since we see those quotes being interpreted and analyzed, the excerpt becomes a secondary source. Choice *C*, tertiary source, is an index of secondary and primary sources, like an encyclopedia or Wikipedia.

17. B: It took two years for the new castle to be built. It states this in the first sentence of the second paragraph. In the third year, we see the Prince planning improvements, and arranging things for the fourth year.

18. C: In this context, *impress* means to impose a certain quality upon. The sentence states that "the impress of his dear hand [has] been stamped everywhere," regarding the quality of his tastes and creations on the house. Choice *A* is one definition of *impress*, but this definition is used more as a verb than a noun: "She impressed us as a songwriter." Choice *B* is incorrect because it is also used as a verb: "He impressed the need for something to be done." Choice *D* is incorrect because it is part of a physical act: "the businessman impressed his mark upon the envelope." The phrase in the passage is figurative, since the workmen did most of the physical labor, not the Prince.

19. C: The foundation of the new house was laid in September of 1853, according to the first paragraph of the text.

20. B: The tone of this passage is excited. We have the author describing something new being built, and the people building the new home are excited for its foundation and its construction.

21. D: *Congealed* in this context most nearly means *thickened*, because we see liquid turning into ice. Choice *B*, *loosened*, is the opposite of the correct answer. Choices *A* and *C*, *burst* and *shrank*, are also incorrect.

22. C: Choice *C* is correct. We cannot infer that the passage takes place during the nighttime. While we do have a statement that says that the darkness thickened, this is the only evidence we have. The darkness could be thickening because it is foggy outside. We don't have enough proof to infer this otherwise. Choice *A* is incorrect; some of the evidence here is that "the cold became intense," and people were decorating their shops with "holly sprigs," a Christmas tradition. It also mentions that it's Christmas time at the end of the passage. Choice *B* is incorrect; we *can* infer that the narrative is located in a bustling city street by the actions in the story. People are running around trying to sell things, the atmosphere is busy, there is a church tolling the hours, etc. The scene switches to the Mayor's house at the end of the passage, but the answer says *majority*, so this is still incorrect. Choice *D* is incorrect; we *can* infer that the Lord Mayor is wealthy—he lives in the "Mansion House" and has fifty cooks.

23. D: Choice *D* is correct because the passage tells us that the poulterers' and grocers' trades were "a glorious pageant, with which it was next to impossible to believe that such dull principles as bargain and sale had anything to do," which means they set up their shops to be entertaining public spectacles to increase sales. Choice *A* is incorrect; although the word *joke* is used, it is meant to be used as a source of amusement rather than something made in poor quality. Choice *B* is incorrect; that they put on a "pageant" is figurative for the public spectacle they made with their shops, not a literal play. Finally, Choice *C* is incorrect; this is not mentioned anywhere in the passage.

24. B: Choice *B* is correct because *proffering* means "to offer something." Choice *A*, giving away, is incorrect. Choice *C* is incorrect because *bolstering* means "helping" or "maintaining." Choice *D*, teaching, is incorrect because it doesn't make sense in the context of the passage.

25. A: Choice *A* is correct. *Ruddy* means "red," so we can deduce that the phrase *made pale faces ruddy* means that the shops made people's faces look red. This is a descriptive sentence, so a careful reading of what's going on is imperative. Choices *B, C,* and *D,* although they may contain components of the original meaning, are incorrect.

Mathematics Achievement

1. B: The fraction $\frac{12}{60}$ can be reduced to $\frac{1}{5}$, in lowest terms. First, it must be converted to a decimal. Dividing 1 by 5 results in 0.2. Then, to convert to a percentage, move the decimal point two units to the right and add the percentage symbol. The result is 20%.

2. B: If a calculator is used, divide 33 into 14 and keep two decimal places. If a calculator is not used, multiply both the numerator and denominator times 3. This results in the fraction $\frac{42}{99}$, and hence a decimal of 0.42.

3. B: Common denominators must be used. The LCD is 15, and $\frac{2}{5} = \frac{6}{15}$. Therefore, $\frac{14}{15} + \frac{6}{15} = \frac{20}{15}$, and in lowest terms, the answer is $\frac{4}{3}$. A common factor of 5 was divided out of both the numerator and denominator.

4. A: A product is found by multiplication. Multiplying two fractions together is easier when common factors are cancelled first to avoid working with larger numbers

$$\frac{5}{14} \times \frac{7}{20}$$

$$\frac{5}{2 \times 7} \times \frac{7}{5 \times 4}$$

$$\frac{1}{2} \times \frac{1}{4} = \frac{1}{8}$$

5. D: Division is completed by multiplying times the reciprocal. Therefore:

$$24 \div \frac{8}{5}$$

$$\frac{24}{1} \times \frac{5}{8}$$

$$\frac{3 \times 8}{1} \times \frac{5}{8}$$

$$\frac{15}{1} = 15$$

6. C: Common denominators must be used. The LCD is 168, so each fraction must be converted to have 168 as the denominator.

$$\frac{5}{24} - \frac{5}{14}$$

$$\frac{5}{24} \times \frac{7}{7} - \frac{5}{14} \times \frac{12}{12}$$

$$\frac{35}{168} - \frac{60}{168} = -\frac{25}{168}$$

7. C: The mean is found by adding all the times together and dividing by the number of times recorded.

$$25 + 18 + 23 + 28 + 30 + 22.5 + 23 + 33 + 20 = 222.5, \text{ divided by } 9 = 24.7$$

Rounding to the nearest minute, the mean is 25.

8. C: The mode is the time from the data set that occurs most often. The number 23 occurs twice in the data set, while all others occur only once, so the mode is 23.

9. A: To find the median of a data set, you must first list the numbers from smallest to largest, and then find the number in the middle. If there are two numbers in the middle, add the two numbers in the middle together and divide by 2. Putting this list in order from smallest to greatest yields 18, 20, 22.5, 23, 23, 25, 28, 30, and 33, where 23 is the middle number, so 23 minutes is the median.

10. D: First, like terms are collected to obtain:

$$12 - 5x = -5x + 12$$

Then, if the addition principle is used to move the terms with the variable, $5x$ is added to both sides and the mathematical statement $12 = 12$ is obtained. This is always true; therefore, all real numbers satisfy the original equation.

11. B: The distributive property is used on both sides to obtain:

$$4x + 20 + 6 = 4x + 6$$

Then, like terms are collected on the left, resulting in:

$$4x + 26 = 4x + 6$$

Next, the addition principle is used to subtract $4x$ from both sides, and this results in the false statement $26 = 6$. Therefore, there is no solution.

12. D: This is a one-step real-world application problem. The unknown quantity is the number of cases of cola to be purchased. Let x be equal to this amount. Because each case costs $3.50, the total number of cases multiplied by $3.50 must equal $40. This translates to the mathematical equation $3.5x = 40$. Divide both sides by 3.5 to obtain $x = 11.4286$, which has been rounded to four decimal places. Because cases are sold whole, and there is not enough money to purchase 12 cases, 11 cases is the correct answer.

13. D: The area of the non-shaded region can be found by adding the areas of the three triangles. The top left triangle has side of 3 and 3, so the formula $A = 1/2 \times \text{base} \times \text{height}$ can be used. This results in an area of 4.5 cm squared. The other two triangles have sides of 3 cm and 6 cm, so the formula can be used again to find an area of 9 cm squared. Since there are two of those same triangles, the sum of the areas is $4.5 + 9 + 9 = 22.5$ cm squared.

14. B: The numbers can be multiplied in a different order to make the product easier to find. 25 multiplied by 4 yields a value of 100. Multiplying 2 times 3 times 10 gives a value of 60. Then 60 times 100 gives a product of 6,000.

15. B: There are 4 triangles and a total of 7 shapes, so the fraction is $\frac{4}{7}$.

16. B: The value 4.9 can be rounded to 5, and the value 3.4 can be rounded to 3. Multiplying 5 times 3 yields a value of 15.

17. B: The fraction of the rectangle that is shaded is $\frac{1}{3}$. There are 2 parts out of 6 that are shaded, and the fraction $\frac{2}{6}$ can be simplified to $\frac{1}{3}$.

18. D: The missing value of x is equal to 9 because $3 + 9 = 12$. The missing value of y is 6 because $4 + 6 = 10$. Adding the missing values of x and y gives an equation of $9 + 6 = 15$.

19. C: The sum of all angles of a triangle is 180 degrees. Since two of the angle measurements are given, the values 100 and 40 can be subtracted from 180 to find a missing angle measure of 40 degrees.

20. C: The correct mathematical statement is the one in which the smaller of the two numbers is on the "less than" side of the inequality symbol. It is written in Choice C that $\frac{1}{3} > -\frac{4}{3}$, which is the same as $-\frac{4}{3} < \frac{1}{3}$, a correct statement.

21. C: $-\frac{1}{5} > \frac{4}{5}$ is incorrect. The expression on the left is negative, which means that it is smaller than the expression on the right. As it is written, the inequality states that the expression on the left is greater than the expression on the right, which is not true.

22. B: A line graph represents continuous change over time. The line on the graph is continuous and not broken, as on a scatter plot. A bar graph may show change but isn't necessarily continuous over time. A pie graph is better for representing percentages of a whole. Histograms are best used in grouping sets of data in bins to show the frequency of a certain variable.

23. C: The volume for a cylinder is:

$$V = \pi \times radius^2 \times height$$

This can be seen here:

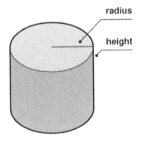

In this case, it would be:

$$V = \pi \times 5^2 \times 10$$

24. A: The chart is a bar chart showing how many men and women prefer each genre of movies. The dark gray bars represent the number of women, while the light gray bars represent the number of men. The light gray bars are higher and represent more men than women for the genres of Comedy and Action.

25. A: First, the distributive property must be used on the left side. This results in:

$$3x + 6 = 14x - 5$$

The addition property is then used to add 5 to both sides, and then to subtract $3x$ from both sides, resulting in $11 = 11x$. Finally, the multiplication property is used to divide each side by 11. Therefore, $x = 1$ is the solution.

26. C: The mean for the number of visitors during the first 4 hours is 14. The mean is found by calculating the average for the four hours. Adding up the total number of visitors during those hours gives:

$$12 + 10 + 18 + 16 = 56$$

Then $56 \div 4 = 14$.

27. C: The mode for a set of data is the value that occurs the most. The grade that appears the most is 95. It's the only value that repeats in the set.

28. A: The area of the shaded region is calculated in a few steps. First, the area of the rectangle is found using the formula:

$$A = length \times width = 6 \text{ m} \times 2 \text{ m} = 12 \text{ m}^2$$

Second, the area of the triangle is found using the formula:

$$A = \frac{1}{2} \times base \times height = \frac{1}{2} \times 3 \text{ m} \times 2 \text{ m} = 3 \text{ m}^2$$

The last step is to take the rectangle area and subtract the triangle area. The area of the shaded region is:

$$A = 12 \text{ m}^2 - 3 \text{ m}^2 = 9 \text{ m}^2$$

29. D: The volume for a cylinder is found by using the formula:

$$V = \pi r^2 h = \pi (2 \text{ in})^2 \times 3.5 \text{ in} = 43.96 \text{ in}^3$$

30. C: There are 0.006 kiloliters in 6 liters because 1 liter is 0.001 kiloliters. The conversion comes from the chart where the prefix kilo- is found three places to the left of the base unit.

Greetings!

First, we would like to give a huge "thank you" for choosing us and this study guide for your ISEE Lower exam. We hope that it will lead you to success on this exam and for years to come.

Our team has tried to make your preparations as thorough as possible by covering all of the topics you should be expected to know. In addition, our writers attempted to create practice questions identical to what you will see on the day of your actual test. We have also included many test-taking strategies to help you learn the material, maintain the knowledge, and take the test with confidence.

We strive for excellence in our products, and if you have any comments or concerns over the quality of something in this study guide, please send us an email so that we can improve.

As you continue forward in life, we would like to remain alongside you with other books and study guides in our library. We are continually producing and updating study guides in several different subjects. If you are looking for something in particular, all of our products are available on Amazon. You can also send us an email!

Sincerely,
APEX Publishing
info@apexprep.com

FREE

Free Study Tips DVD

In addition to the tips and content in this guide, we have created a FREE DVD with helpful study tips to further assist your exam preparation. **This FREE Study Tips DVD provides you with top-notch tips to conquer your exam and reach your goals.**

Our simple request in exchange for the strategy-packed DVD is that you email us your feedback about our study guide. We would love to hear what you thought about the guide, and we welcome any and all feedback—positive, negative, or neutral. It is our #1 goal to provide you with top quality products and customer service.

To receive your **FREE Study Tips DVD**, email freedvd@apexprep.com. Please put "FREE DVD" in the subject line and put the following in the email:

> a. The name of the study guide you purchased.
>
> b. Your rating of the study guide on a scale of 1-5, with 5 being the highest score.
>
> c. Any thoughts or feedback about your study guide.
>
> d. Your first and last name and your mailing address, so we know where to send your free DVD!

Thank you!